MyWorkbook
with Chapter Summaries

Denise Heban

Basic College Mathematics
through Applications
Fifth Edition

Geoffrey Akst

Borough of Manhattan Community College, The City University of New York

Sadie Bragg

Borough of Manhattan Community College, The City University of New York

PEARSON

Boston Columbus Indianapolis New York San Francisco Upper Saddle River
Amsterdam Cape Town Dubai London Madrid Milan Munich Paris Montreal Toronto
Delhi Mexico City Sao Paulo Sydney Hong Kong Seoul Singapore Taipei Tokyo

The author and publisher of this book have used their best efforts in preparing this book. These efforts include the development, research, and testing of the theories and programs to determine their effectiveness. The author and publisher make no warranty of any kind, expressed or implied, with regard to these programs or the documentation contained in this book. The author and publisher shall not be liable in any event for incidental or consequential damages in connection with, or arising out of, the furnishing, performance, or use of these programs.

Reproduced by Pearson from electronic files supplied by the author.

ISBN-13: 978-0-321-75977-1
ISBN-10: 0-321-75977-X

1 2 3 4 5 6 OPM 15 14 13 12 11

www.pearsonhighered.com

PEARSON

CONTENTS

Name: Date:
Instructor: Section:

Chapter 1 WHOLE NUMBERS

1.1 Introduction to Whole Numbers

Objectives
A   Read and write whole numbers.
B   Write whole numbers in expanded form.
C   Round whole numbers.
D   Solve applied problems involving reading, writing, or rounding whole numbers.

MATHEMATICALLY SPEAKING
In exercises 1-7 fill in the blank with the most appropriate term or phrase from the given list.

even	placeholder	rounded	digits
place value	whole numbers	odd	expanded form
standard form			

1.   The _____ are 0, 1, 2, 3, 4, 5, … .

2.   The whole numbers are written with the _____ 0, 1, 2, 3, 4, 5, 6, 7, 8, and 9.

3.   The numbers 1, 3, 5, 7, 9, … are _____.

4.   When the number 279 is written as 2 hundreds + 7 tens + 9 ones, it is said to be in _____.

5.   The number 636 _____ to the nearest ten is 640.

6.   The number eighty-six, when written as 86, is said to be in _____.

7.   In the number 342, the _____ of the 2 is ones.

EXAMPLES AND PRACTICE

Review this example for Objective A:

Read and write whole numbers.

1. In each number, identify the place that the digit 8 occupies.

   a. 218
      The ones place

   b. 984,316
      The ten thousands place

2. How do you read the number 4,017?

   "four thousand, seventeen"

3. Write the number ten million, seventy-eight thousand in standard form.

   10,078,000

Review this example for Objective B:

Write whole numbers in expanded form.

4. Write in expanded form 3,017.

Thousands	Hundreds	Tens	Ones
3	0	1	7

   3,017 is 3 thousands + 0 hundreds + 1 ten + 7 ones or 3,000 + 10 + 7

Review this example for Objective C:

Round whole numbers.

5. Round 65,728 to the nearest thousand

   65,728 = 6<u>5</u>,728 Underline the digit in the thousands place

   = 65,728 The critical digit, 7, is greater than 5; add 1 to the underlined digit

   ≈ 66,000 Change the digits to the right of the underlined digit to 0's.

Practice:

1. In each number, identify the place that the digit 6 occupies.

   a. 3,016

   b. 456,982

2. How do you read the number 56,210?

3. Write the number twenty million, sixty thousand in standard form.

4. Write 21,459 in expanded form.

5. Round 57,920 to the nearest thousand

Review this example for Objective D:

Solve applied problems involving reading, writing, or rounding whole numbers.

6. One of the largest giant sequoias in the United States is 3,288 feet tall. Round this number to the nearest thousand.

$3,288 = \underline{3},288$ Underline the digit in the thousands place

$= \underline{3},288$ The critical digit, 2, is less than 5; do not add 1

$\approx 3,000$

6. A movie theatre recently underwent renovations. The company spent $239,305 in renovations. How much is this to the nearest ten thousand?

ADDITIONAL EXERCISES

Objective A Read and write whole numbers.
Insert commas as needed.
1. 8 9 0 0 3

2. 3 2 6 5 9 0

Write each number in words.
3. 23,890

4. 12,053,007

Identify the place occupied by the underline digit.
5. $\underline{5}2,841$

6. $87\underline{3},911$

Write each number in standard form.
7. Forty-six thousand, two hundred nine

8. Twelve billion, seven hundred million, one hundred ninety-two

Objective B Write whole numbers in expanded form.
Write each number in expanded form.

9. 986

10. 3,017

Objective C Round whole numbers
Round to the indicated place.

11. 12,752 to the nearest hundred   **12.** 268,945 to its largest place

Objective D Solve applied problems involving reading, writing, or rounding whole numbers.

Solve.

13. A roller coaster is 161 feet tall. What is the height rounded to the nearest ten?

14. The area of Indonesia is 705,189 square miles. Round the area to the nearest thousand.

Chapter 1 WHOLE NUMBERS

1.2 Adding and Subtracting Whole Numbers

Objectives
A   Add or subtract whole numbers.
B   Solve applied problems involving the addition or subtraction of whole numbers.

MATHEMATICALLY SPEAKING

In exercises 1-6 fill in the blank with the most appropriate term or phrase from the given list.

subtrahend	**Commutative Property of Addition**	**addends**	**sum**
difference	**Identity Property of Addition**		
estimates	**Associative Property of Addition**		

1. The _____ states that the sum of a number and zero is the original number.

2. In an addition problem the numbers being added are called _____.

3. The _____ states that changing the order in which two numbers are added does not affect the sum.

4. In a subtraction problem, the number being subtracted is called the _____.

5. The _____ states that when adding three numbers, regrouping addends gives the same sum.

6. The result of subtraction is called the _____.

EXAMPLES AND PRACTICE

| **Review this example for Objective A:** | **Practice:** |

Add or subtract whole numbers.

1. Add 9,621 and 512.
Write the problem vertically with the addends lined up on the right.

```
  9 6 2 1
+   5 1 2
```
       3   Add the ones

```
  9 6 2 1
+   5 1 2
```
     3 3   Add the tens

```
1
  9 6 2 1
+   5 1 2
10 1 3 3
```
   Add the hundreds; We write the 1 and carry the 1 to the thousands column.

---

1. Add 46,319 and 10,520.

---

2.   Subtract

```
  513
−  92
```
     1   subtract the ones

```
  4 11
  5̸ 1̸ 3
−   9 2
  421
```
   subtract the tens; borrow from the hundreds

---

2. Subtract

```
  741
−350
```

---

3. Compute the sum of 2,398 + 745 + 412 + 1,876. Check by estimation.

   2,398
    745
    412
+ 1,876
  5,431 Exact sum

Check We round to the largest place value.

$$\begin{array}{rcr}
2,398 & \approx & 2,000 \\
745 & \approx & 700 \\
412 & \approx & 400 \\
+\ 1,876 & \approx & +\ 2,000 \\
\hline
& & 5,100 \ \text{Estimated sum}
\end{array}$$

3. Compute the sum of 14,325 + 10,459 + 863. Check by estimation.

---

4. Subtract 1,884 from 8,329. Check by estimating the difference.

  8,329
− 1,884
  6,445 Exact difference

Check Round to the nearest thousand.

$$\begin{array}{rcr}
8,329 & \approx & 8,000 \\
-\ 1,884 & \approx & -\ 2,000 \\
\hline
& & 6,000 \ \text{Estimated difference}
\end{array}$$

4. Subtract: 6,932 − 3,105. Check by estimating the difference.

Review this example for Objective B:

Solve applied problems involving the addition or subtraction of whole numbers.

5. The current balance in Cassie's bank account is $739. She deposits her paycheck of $1,238. What is the new balance?

To find the new balance add: 1,238 + 739.

  1,238
+   739
  1,977

The new balance is $1,977.

5. The current balance in Maria's bank account is $4,739. She deposits her paycheck of $867. What is the new balance?

6. The current balance in Cassie's bank account is $1,465. She withdraws $246. What is the new balance?

To find the new balance, subtract:
1,465 – 246

$$
\begin{array}{r}
5\ 15\\
1,\ 4\ \cancel{6}\ \cancel{5}\\
-\quad 2\ 4\ 6\\
\hline
1,\ 2\ \ 1\ 9
\end{array}
$$

The new balance is $1,219.

6. The current balance in Maria's bank account is $1,297. She withdraws $439. What is the new balance?

ADDITIONAL EXERCISES

Objective A Add or subtract whole numbers.

Add or subtract.

1.   4,663
       + 371

2.   81,452
       +25,199

3.   46,319
       +10,520

4.   3,750
            1,725
          +4,992

5.   3,227
            2,806
          +5,481

6.   90,316
           10,882
          + 5,281

7. $281 + 758 + 104 + 533$

8. $\$845 + \$39 + \$1,871 + \$2,755$

9. 304
          -129
          $\overline{}$

10. $982,111$
           $-613,101$
           $\overline{}$

11. $8,286 - 3,100$

12. $12,799 - 2,357$

Estimate the sum or difference

13. $14,723 + 8,211$

14. $227,410 - 130,817$

Objective B Solve applied problems involving the addition or subtraction of whole numbers.

15. In 1990, the population of Columbus, Ohio was 1,405,000. During the next ten years, the population grew by 208,000 people. What was the population in 2000?

16. You are driving along a highway whose speed limit is 70 mph. Your speed is 57 mph. How much faster can you drive without going over the speed limit?

17. Your dental bill was $800. Your dental insurance reimbursed your for $350. How much of this bill was not reimbursed?

Chapter 1 WHOLE NUMBERS

1.3 Multiplying Whole Numbers

Objectives
A   Multiply whole numbers.
B   Solve applied problems involving the multiplication of whole numbers.

MATHEMATICALLY SPEAKING
In exercises 1-4 fill in the blank with the most appropriate term or phrase from the given list.

> **Identity Property of Multiplication   subtraction   Multiplication Property of 0**
> **addition   Distributive Property   sum   product**

1. The _____ is illustrated by $5 \times (4+9) = (5 \times 4) + (5 \times 9)$.

2. The multiplication of whole numbers can be thought of as repeated _____.

3. The _____ states that the product of any number and 1 is that number.

4. The result of multiplying two factors is called their _____.

Name: Date:

Instructor: Section:

EXAMPLES AND PRACTICE

Review this example for Objective A:	Practice:
Multiply whole numbers.	

1. Multiply: $504 \cdot 8$
 Write the problem vertically.
 	
      3
    504
 $\underline{\times \quad 8}$
   $4,032$
	1. Multiply: $704 \cdot 6$

2.   Simplify: $459 \cdot 208$
 	
      459
 $\underline{\times \quad 208}$
     3672 $\leftarrow 8 \times 459$
      000     $\leftarrow 0 \times 459$
   $\underline{91800}$ $\leftarrow 2 \times 459$
   $95,472$
	2. Simplify: $367 \cdot 408$

3.   Multiply 418 by 186. Check the answer by estimation.
      418
 $\underline{\times \quad 186}$
     2508
   33440
 $\underline{41800}$
 $77,748 \leftarrow$ Exact product
 Check:
    418
 $\underline{\times \quad 186}$
   $77,748$
 $\approx$ $400 \leftarrow$ the largest place is hundreds
 $\approx \underline{\times \quad 200} \leftarrow$ the largest place is hundreds
   $80,000 \leftarrow$ estimated product
	3. Multiply 528 by 146. Check the answer by estimation.

Review this example for Objective B:

Solve applied problems involving the multiplication of whole numbers.

4. The textbook for a course in government costs $112. There are 42 students in the class. What is the total cost of the textbooks?

$$112$$
$$\times\ 42$$
$$\overline{224}$$
$$\underline{4480}$$
$$4704$$

The total cost of the textbooks is $4,704.

4. A one ounce (28 g) serving of peanuts contains 160 calories. How many calories are in a 6 ounce serving?

ADDITIONAL EXERCISES

Objective A Multiply whole numbers.

Multiply and check.

1. 398
    × 4

2. 8,381
    ×   5

3. 892
    × 35

4. 992
    × 68

5. 4,037
    × 26

6. 2,349
    × 32

7.   3,429
      ×   234

8.   16,051
      × 2,413

Estimate the product.

9.   (405)(31)

10. 682×39

11. $4,723 \times 23$

12. $58 \times 82 \times 5$

Objective B Solve applied problems involving the multiplication of whole numbers.

13. Your car gets 28 miles per gallon (mpg) of gasoline. If a full tank contains 18 gallons, how many miles can you drive?

14. Albany, NY is 655 miles from Detroit, MI. Boise, ID is about three times as far from Detroit as Albany. How far is Boise from Detroit?

15. The monthly rent for an apartment is $860. How much is paid in rent for an entire year?

16. Janice can type 35 words a minute. Approximately how many words can she type in 45 minutes?

Chapter 1 WHOLE NUMBERS

1.4 Dividing Whole Numbers

Objectives
A   Divide whole numbers.
B   Solve applied problems involving the division of whole numbers.

MATHEMATICALLY SPEAKING

In exercises 1-4 fill in the blank with the most appropriate term or phrase from the given list.

subtraction   quotient   divided   product   divisor   increased   multiplication

1.   The result of dividing is called the _____.

2.   Any whole number _____ by 1 is equal to the number itself.

3.   When dividing, the dividend is divided by the _____.

4.   The opposite operation of division is _____.

EXAMPLES AND PRACTICE

Review this example for Objective A:	**Practice:**
Divide whole numbers.	

1. Divide and check: $5,632 \div 8$

$$\begin{array}{r} 704 \\ 8\overline{)5632} \end{array}$$

$\underline{56}$ $\quad$ 8(7) = 56. Subtract.

03    $\quad$ There are zero 8's in 3.

$\underline{\;0}$ $\quad$ 0(8) = 0. Subtract.

32

$\underline{32}$ $\quad$ 4(8) =32. Subtract.

0

Check:

704

$\underline{\times\;\;8}$

$5,632$

1. Divide and check: $4,572 \div 9$

---

2. Compute: $\dfrac{2,237}{31}$ and check.

$$\begin{array}{r} 72 \\ 31\overline{)2237} \end{array}$$

$\underline{217}$

67

$\underline{62}$

5

Check: $31 \times 72 + 5$

$\quad\quad 2,232 + 5 = 2,237$

2. Compute: $\dfrac{2,920}{36}$ and check.

3. Calculate $\dfrac{8,008}{44}$ and then check by estimation.

$$
\begin{array}{r}
182 \\
44\overline{)8008} \\
\underline{44} \\
360 \\
\underline{352} \\
88 \\
\underline{88} \\
0
\end{array}
$$

Check: Round 8,008 to 8,000
Round 44 to 40
$8000 \div 40 = 200$

3. Calculate $\dfrac{1,768}{34}$ and then check by estimation.

Review this example for Objective B:

Solve applied problems involving the division of whole numbers.

4. Four college students share a paper route to earn money for textbooks. The total in their fund was $3,224. How much is each student's equal share?

$$
\begin{array}{r}
806 \\
4\overline{)3224} \\
\underline{32} \\
02 \\
\underline{0} \\
24 \\
\underline{24} \\
0
\end{array}
$$

Each student receives $806.

4. Three friends go together and purchase 171 blank DVD's to share equally. How many DVD's does each friend receive?

ADDITIONAL EXERCISES

Objective A Divide whole numbers.

Divide and check.

1. $3\overline{)150}$ **2.** $2\overline{)482}$

3. $6\overline{)726}$ **4.** $4\overline{)904}$

5. $\dfrac{750}{50}$ **6.** $\dfrac{2,400}{30}$

7. $10,179 \div 87$ **8.** $99,904 \div 33$

Estimate the quotient.

9. $352,127 \div 65$ **10.** $17,382 \div 56$

11. $23,938 \div 389$ **12.** $121,475 \div 261$

Objective B Solve applied problems involving the division of whole numbers.

13. A family spent $436 for four nights in a hotel. How much money did they spend per night?

14. Your new laser printer prints 21 pages per minute. How long will it take to print your 189-page report?

15. Bejnar purchased a pay-as-you-go phone card for $30.00. If each call cost a flat fee of 8¢ per minute, how many phone calls can Bejnar make?

16. Thomas can type approximately 45 words per minute. How long will it take him to type a 900 word essay for his English class?

Chapter 1 WHOLE NUMBERS

1.5 Exponents, Order of Operations, and Averages

Objectives
A   Evaluate expressions involving exponents.
B   Evaluate expressions using the rule for order of operations.
C   Compute averages.
D   Solve applied problems involving exponents, order of operations, or averages.

MATHEMATICALLY SPEAKING
In exercises 1-4 fill in the blank with the most appropriate term or phrase from the given list.

product   subtracting   sum   grouping   adding   power   listing   base

1.   In evaluating an expression involving both a sum and a product, the
     _____ is evaluated first.

2.   An average of values on a list is found by _____ the values and then
     dividing by how many values there are on the list.

3.   Parentheses and brackets are examples of _____ symbols.

4.   An exponent indicates how many times the _____ is multiplied by itself.

EXAMPLES AND PRACTICE

Review this example for Objective A:	**Practice:**
Evaluate expressions involving exponents.	
1. Rewrite $2 \cdot 2 \cdot 3 \cdot 3 \cdot 3$ in exponential notation. $$\underbrace{2 \cdot 2}_{\text{2 factors of 2}} \cdot \underbrace{3 \cdot 3 \cdot 3}_{\text{3 factors of 3}} = 2^2 \cdot 3^3$$	**1.** Rewrite $4 \cdot 4 \cdot 4 \cdot 5 \cdot 5$ in exponential notation.
2. Write $3^3 \cdot 2^4$ in standard form. $$\begin{aligned} 3^3 \cdot 2^4 &= (3 \cdot 3 \cdot 3) \cdot (2 \cdot 2 \cdot 2 \cdot 2) \\ &= 27 \cdot 16 \\ &= 432 \end{aligned}$$	**2.** Write $5^3 \cdot 2^5$ in standard form.
Review this example for Objective B:	
Evaluate expressions using the rule for order of operations.	
3. Evaluate: $15 - 8 \div 2$ $$15 - \underline{8 \div 2} =$$ $$15 - 4 = 11$$	**3.** Evaluate: $7 \times (2 + 3) - 21$
4. $6 + 4 \cdot (7 + 4^2) = 6 + 4 \cdot \underbrace{(7 + 16)}$ $$\begin{aligned} &= 6 + \underline{4 \cdot 23} \\ &= \underline{6 + 92} \\ &= 98 \end{aligned}$$	**4.** Evaluate: $8 + 5 \cdot (2 + 5^2)$

Review this example for Objective C:

Compute averages.

5. What is the average of 25, 32, 47, and 56?

The average equals the sum of the numbers divided by 4.

$$\frac{25 + 32 + 47 + 56}{4} = \frac{160}{4} = 40$$

5. What is the average of 6, 9, 12, 15, and 18?

Review this example for Objective D:

Solve applied problems involving exponents, order of operations, or averages.

6. The high temperatures for seven days were: 79°, 83°, 88°, 73°, 75°, 75°, 80°. What was the average temperature?

$$\frac{\text{daily high temperatures}}{\text{number of days}}$$

$$= \frac{79 + 83 + 88 + 73 + 75 + 75 + 80}{7}$$

$$= \frac{553}{7}$$

$$= 79°$$

6. A college basketball team scored 85, 93, 80, 85, 92, and 105 in its first six games. What was the teams average points per game (ppg)?

Name: Date:

Instructor: Section:

ADDITIONAL EXERCISES

Objective A Evaluate expressions involving exponents.

Write in exponential form. Do not evaluate.

1. $9 \cdot 4 \cdot 9 \cdot 4 \cdot 4 \cdot 9 \cdot 9$

2. $2 \cdot 2 \cdot 2 \cdot 2 \cdot 11 \cdot 11 \cdot 11 \cdot 11 \cdot 11$

3. $2 \cdot 2 \cdot 2 \cdot 7 \cdot 7 \cdot 7 \cdot 7 \cdot 13 \cdot 13$

4. $5 \cdot 11 \cdot 11 \cdot 17 \cdot 17 \cdot 17$

Write each number in standard form.

5. $5^2 \cdot 3^3$

6. $8^2 \cdot 10^3$

Objective B Evaluate expressions using the rule for order of operations.

Evaluate.

7. $9^2 - (20 + 11)$

8. $6 \times 4 + 5^2 - 11$

9. $8 \times 9 - 6^2 \div 4$

10. $5^2 - 4^2 + 3 \times 2$

Objective C Compute averages.

Find the average of each set of numbers.

11.   29, 29, 29, 27, 25, and 23

12.   79, 64, 69, 79, and 84

Objective D Solve applied problems involving exponents, order of operations, or averages.

13. In seven games, Jeremy bowled 163, 185, 154, 127, 156, 140, and 139. What was his average score for these games?

14. Your test scores on the four exams before the final exam were 88, 72, 82, and 94. What was your average test score before the final exam?

Chapter 1 WHOLE NUMBERS

1.6 More on Solving Word Problems

Objective

A   Solve applied problems involving the addition, subtraction, multiplication, or division of whole numbers using various problem-solving strategies.

MATHEMATICALLY SPEAKING

In exercises 1-4 fill in the blank with the most appropriate term or phrase from the given list.

   read   check   clue words   problem-solving strategy   operations

1. When solving a word problem, it is important to first _____ the problem carefully.

2. The last step when solving a problem is to _____ the solution.

3. Drawing a picture, substituting simpler numbers, or making a table are all examples of a _____.

4. Words that suggest performing a particular mathematical operation are called _____.

EXAMPLES AND PRACTICE

Review this example for Objective A:	**Practice:**
Solve applied problems involving the addition, subtraction, multiplication, or division of whole numbers using various problem-solving strategies.	
1. The John Hancock tower in Chicago stands 1,127 feet tall, while the Willis Tower is 1,450 feet tall. What is the difference between the heights of these two buildings? You are looking for a difference. Subtract the heights. $1,450 - 1,127 = 323$ The difference in the heights is 323 feet.	1. A local mining company extracted 1,525 tons of coal in January. In March, they extracted 1,832 tons. How many more tons were extracted in March?
2. You are the manager of a local outdoor concert stadium. A total of 3,375 tickets have been sold for a concert. Your plan is to have 75 equal rows of seats. How many seats will be needed per row? You àre looking for the number of seats needed per row. Divide the number of tickets sold by the number of rows. $\dfrac{3,375}{75} = 45$ You will need 45 seats per row.	2. A local business man has donated $7,650 to be divided equally among 6 schools. How much money will each school receive?
3. Sandra pays $675 per month in rent. How much does she pay in a year? Multiply the monthly rent by 12 months in a year. $\$675 \cdot 12 = \$8,100$ Sandra pays $8,100 for the year.	3. Carlos has a part-time job that pays $236 per week. How much will Carlos make working 26 weeks?

Name: _____

Date: _____

Instructor: _____

Section: _____

ADDITIONAL EXERCISES
Objective A Solve applied problems involving the addition, subtraction, multiplication, or division of whole numbers using various problem-solving strategies.

Solve and check.

1. Last year, food expenses amounted to $7,228. This year, they were $9,055. How much of an increase was this?

2. If you type 55 words per minute, how long will it take you to type an 11-page report with about 400 words on each page?

3. It took 27 weeks for the work crew to build a road 8,316 feet in length. How many feet of road did the crew build per week?

4. The walls of your heart are composed of muscles that contract about 100,000 times a day. About how many contractions are there in a 30-day month?

5. A college library contains 84,601 books. How many books is this to the nearest ten thousand?

6. Cherie bought a 12-foot by 12-foot rug for her living room. The living room measures 15 feet by 15 feet. How much of the floor will be exposed after the rug is laid?

7. Juan is saving to buy a $175 portable DVD player. During the past three months he has saved $57, $26, and $34. How much more money does he need?

8. The total fall enrollment in private degree-granting institutions in Pennsylvania from 2008 to 2012 is given in the table. What was the average fall enrollment in the years 2008 through 2012? Round to the nearest whole number.

Year	Enrollment
2008	270,292
2009	276,349
2010	284,440
2011	294,320
2012	304,255

9. Your senior class is taking a trip and needs to rent some busses. A bus can seat 40 passengers. If a total of 248 people are taking the trip, how many busses must be rented?

10. A woman who weighed 208 pounds went on a diet and now weighs 124 pounds. She took 14 months to lose the weight. How much, on the average, did she lose each month?

Chapter 2 FRACTIONS

2.1 Factors and Prime Numbers

Objectives
A   Find the factors of a whole number.
B   Identify prime and composite numbers.
C   Find the prime factorization of a whole number.
D   Find the least common multiple of two or more numbers.
E   Solve applied problems using factoring or the LCM.

MATHEMATICALLY SPEAKING
In exercises 1-5 fill in the blank with the most appropriate term or phrase from the given list.

factors   least common multiple   prime factorization   common multiple
prime   composite   factor tree

1.   A number that is a multiple of two or more numbers is a _____
     of these numbers.

2.   A number written as the product of its prime factors is called its _____.

3.   A(n) _____ number is a whole number that has more than two
     factors.

4.   The _____ of two or more numbers is the smallest
     nonzero number that is a multiple of each number.

5.   The numbers 1, 2, 3, 4, 6, 8, 12, and 24 are _____ of 24.

EXAMPLES AND PRACTICE

| Review this example for Objective A: | Practice: |

Find the factors of a whole number.

1. Find all the factors of 8.

$$\frac{8}{1} = 8R0 \qquad \frac{8}{2} = 4R0 \qquad \frac{8}{3} = 2R2$$

$$\frac{8}{4} = 2R0 \qquad \frac{8}{5} = 1R3 \qquad \frac{8}{6} = 1R2$$

$$\frac{8}{7} = 1R1 \qquad \frac{8}{8} = 1R0$$

1, 2, 4 and 8 are factors of 8.

1. Find all the factors of 10.

Review this example for Objective B:

Identify prime and composite numbers.

2. a. 29

    b. 44

    c. 18

    a. 29 has only two factors: 1 and 29; prime
    b. 44 is even, meaning 2 is a factor; composite
    c. 18 is even, meaning 2 is a factor; composite

2. Identify whether each number is prime or composite.

    a. 100

    b. 43

    c. 36

Name: Date:
Instructor: Section:

Review this example for Objective C:

Find the prime factorization of a whole number.

3. Express 90 as the product of prime factors.

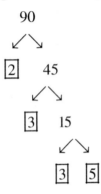

The prime factorization of 90 is
$2 \times 3 \times 3 \times 5$ or $2 \cdot 3^2 \cdot 5$.

---

3. Express 42 as the product of prime factors.

Review this example for Objective D:

Find the least common multiple of two or more numbers.

4. Find the LCM of 12 and 16.

Write the prime factorization of each number.

$12 = 2^2 \cdot 3$

$16 = 2^4$

LCM $= 2^4 \cdot 3 = 16 \cdot 3 = 48$

---

4. Find the LCM of 15 and 18.

Review this example for Objective E:	
Solve applied problems using factoring or the LCM.	**5.** As part of a store opening promotion, every sixth customer receives a red ball and every eighth customer receives a water bottle. Which customer will receive both items?
5. Tina walks a poodle every fourth day. She walks a boxer every sixth day. On what day will Tina walk both dogs?	

$4 = 2^2$

$6 = 2 \cdot 3$

LCM $= 2^2 \cdot 3 = 12$

Every 12th day she will walk both dogs.

ADDITIONAL EXERCISES

Objective A Find the factors of a whole number.
Identify all the factors of each number.

1. 48   **2.** 63

3. 36   **4.** 102

Objective B Identify prime and composite numbers.
Indicate whether each number is prime or composite. If it is composite, identify a factor other than the number itself and 1.

5. 46   **6.** 53

7. 243   **8.** 57

Objective C Find the prime factorization of a whole number.
Write the prime factorization of each number.

9. 36   **10.** 216

11. 252   **12.** 525

Objective D Find the least common multiple of two or more numbers.
Find the LCM of each set of numbers.

13. 4 and 14   **14.** 9 and 15

15. 12 and 21   **16.** 7 and 42

17. 4, 6, and 8   **18.** 3, 7, and 9

19. 5, 8, and 10   **20.** 6, 12, and 52

Objective E Solve applied problems using factoring or the LCM.
Solve.

21. In 2004, the Summer Olympic Games was held in Athens, Greece. If the Summer Olympics are held every four years, will there be Summer Olympics scheduled for 2052?

22. Hakim can run 6 laps around a track in the same time that Kohl can run 4 laps around the same track. After how many laps will they both be at the starting point on the track?

23. Jeff and Cindy both work at General Motors. Cindy gets a day off every 4 days and Jeff gets a day off every 5 days. If they are both off today, in how many days will they both be off again?

24. The planets revolve around the sun. Saturn takes 30 years and Uranus 84 years to make a complete revolution. In how many years will the planets line up in the same way as they do tonight?

25. The planets revolve around the sun. Saturn takes 30 years, Uranus 84 years, and Neptune 165 years to make a complete revolution. In how many years will the planets line up in the same way as they do tonight?

Chapter 2 FRACTIONS

2.2 Introduction to Fractions

Objectives
A   Read or write fractions or mixed numbers.
B   Write improper fractions as mixed numbers or mixed numbers as improper fractions.
C   Find equivalent fractions or write fractions in simplest form.
D   Compare fractions.
E   Solve applied problems with fractions.

MATHEMATICALLY SPEAKING

In exercises 1-2 fill in the blank with the most appropriate term or phrase from the given list.

multiply   divide   simplify

1.   To _____ rational expressions, multiply the numerators and multiply the denominators.

2.   To _____ rational expressions, multiply the first rational expression by the reciprocal of the second rational expression.

EXAMPLES AND PRACTICE

Review this example for Objective A:

Read or write fractions or mixed numbers.

1. In the diagram what does the shaded portion represent?

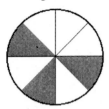

The whole is divided into 8 equal parts, so the denominator of the fraction is 8.

Three of the parts are shaded, so the numerator is 3.

The diagram represents $\frac{3}{8}$.

Practice:

1. In the diagram what does the shaded portion represent?

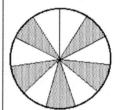

Review this example for Objective B:

Write improper fractions as mixed numbers and mixed numbers as improper fractions.

2. Write $4\frac{3}{7}$ as an improper fraction.

$$4\frac{3}{7} = \frac{(7 \times 4) + 3}{7}$$
$$= \frac{28 + 3}{7} = \frac{31}{7}$$

2. Write $5\frac{2}{9}$ as an improper fraction.

---

3. Write $\frac{13}{3}$ as a mixed or whole number.

$$\frac{13}{3} = 3\overline{)13}^{\,4\,R1}$$

$$\frac{13}{3} = 4\frac{1}{3}$$

3. Write $\frac{15}{4}$ as a mixed or whole number.

Review this example for Objective C:

Find equivalent fractions or write fractions in simplest form.

4. Find two fractions equivalent to $\frac{4}{5}$.

Multiply the numerator and denominator by 4 and then by 9.

$$\frac{4}{5} = \frac{4\cdot 4}{5\cdot 4} = \frac{16}{20} \qquad \frac{4}{5} = \frac{4\cdot 9}{5\cdot 9} = \frac{36}{45}$$

4. Find two fractions equivalent to $\frac{6}{7}$.

5. Write $\frac{28}{50}$ in lowest terms.

$$\frac{28}{50} = \frac{2\cdot 2\cdot 7}{2\cdot 5\cdot 5}$$

$$= \frac{\cancel{2}\cdot 2\cdot 7}{\cancel{2}\cdot 5\cdot 5}$$

$$= \frac{14}{25}$$

5. Write $\frac{30}{55}$ in lowest terms.

Review this example for Objective D:

Compare fractions.

6. Compare $\frac{7}{10}$ and $\frac{5}{8}$.

Write each fraction with a common denominator.

$$\frac{7}{10} = \frac{7\cdot 8}{10\cdot 8} = \frac{56}{80} \qquad \frac{5}{8} = \frac{5\cdot 10}{8\cdot 10} = \frac{50}{80}$$

Compare the numerators. $56 > 50$

$$\frac{7}{10} > \frac{5}{8}$$

6. Compare $\frac{3}{4}$ and $\frac{11}{15}$.

Review this example for Objective E:

Solve applied problems with fractions.

7. If 220 out of 380 students in a lecture
   class are males, what fraction of the
   students are females?

   We know 220 out of 380 are male.
   We need to know how many are female.
   380 – 220 = 160 are female
   $$\frac{160}{380} = \frac{160 \div 20}{380 \div 20} = \frac{8}{19}$$

7. In a shipment of 50 cartons of
   books, a bookstore received
   32 on time, 8 were late and the
   rest never arrived. What
   fraction of the cartons did not
   arrive?

ADDITIONAL EXERCISES

Objective A Read or write fractions or mixed numbers.

Identify a fraction or mixed number represented by the shaded portion.

1.

2.

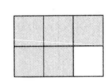

Indicate whether each number is a proper fraction, an improper fraction, or a mixed number.

3. $\frac{12}{7}$

4. $\frac{18}{18}$

5. $33\frac{1}{3}$

Objective B Write improper fractions as mixed numbers or mixed numbers as improper fractions.

Write each number as an improper fraction.

6. 25

7. $3\frac{5}{9}$

8. $12\frac{4}{5}$

Express each fraction as a mixed or whole number.

9. $\dfrac{17}{7}$ **10.** $\dfrac{58}{9}$ **11.** $\dfrac{72}{9}$

Objective C Find equivalent fractions or write fractions in simplest form.

Write as an equivalent fraction with indicated denominator.

12. $\dfrac{5}{9} = \dfrac{}{54}$ **13.** $\dfrac{2}{3} = \dfrac{}{39}$

Simplify, if possible.

14. $\dfrac{8}{12}$ **15.** $\dfrac{15}{33}$

16. $\dfrac{75}{250}$ **17.** $\dfrac{65}{20}$

Objective D Compare fractions.

Between each pair of numbers, insert the appropriate sign: <, =, or >.

18. $\dfrac{2}{7}$ $\dfrac{3}{8}$ **19.** $3\dfrac{6}{11}$ $3\dfrac{5}{12}$

Objective E Solve applied problems with fractions.

Solve. Write your answers in simplest form.

20. A math class has 14 women and 17 men. What is the fraction of women in the class?

21. Matt walked $\dfrac{5}{6}$ mile to a friend's house, and then $\dfrac{3}{4}$ mile to work. Which of the two distances walked is the least?

22. A cubic meter of concrete mix contains 420 kg of cement, 150 kg of stone, and 120 kg of sand. What fractional part of the mix is sand?

Chapter 2 FRACTIONS

2.3 Adding and Subtracting Fractions

Learning Objectives
A   Add or subtract fractions or mixed numbers.
B   Solve applied problems involving the addition or subtraction of fractions or mixed numbers.

MATHEMATICALLY SPEAKING

In exercises 1-3 fill in the blank with the most appropriate term or phrase from the given list.

improper   equivalent   numerators   borrow   subtract   denominators

1. When subtracting $3\frac{4}{5}$ from $6\frac{2}{5}$, _____ from the 6 on the top.

2. To add like fractions, add the _____ and keep the same denominator.

3. To subtract unlike fractions, rewrite them as _____ fractions with the same denominators.

Name: Date:

Instructor: Section:

EXAMPLES AND PRACTICE

Review this example for Objective A:

Add or subtract fractions or mixed numbers.

Practice:

1. Add: $\dfrac{8}{15} + \dfrac{4}{15}$

1. Add: $\dfrac{9}{20} + \dfrac{5}{20}$

$$\dfrac{8}{15} + \dfrac{4}{15} = \dfrac{8+4}{15} = \dfrac{12}{15}$$

$$= \dfrac{2 \cdot 2 \cdot 3}{3 \cdot 5} = \dfrac{2 \cdot 2 \cdot \cancel{3}}{\cancel{3} \cdot 5} = \dfrac{4}{5}$$

---

2. Find the difference between $\dfrac{13}{5}$ and $\dfrac{2}{5}$.

$$\dfrac{13}{5} - \dfrac{2}{5} = \dfrac{13-2}{5} = \dfrac{11}{5} \text{ or } 2\dfrac{1}{5}$$

2. Find the difference between $\dfrac{15}{6}$ and $\dfrac{4}{6}$.

---

3. Add: $\dfrac{4}{15} + \dfrac{17}{35}$

Find the LCD, then find equivalent fractions and add the numerators, keeping the same denominator.

$$\dfrac{4}{15} = \dfrac{4 \cdot 7}{15 \cdot 7} = \dfrac{28}{105}$$

$$+ \dfrac{17}{35} = \dfrac{17 \cdot 3}{35 \cdot 3} = \dfrac{51}{105}$$

$$= \dfrac{79}{105}$$

3. Add: $\dfrac{5}{6} + \dfrac{3}{4}$

---

4. Add: $8\dfrac{3}{4} + 9\dfrac{5}{6}$

The LCD is 12.

$$8\dfrac{3}{4} = 8\dfrac{9}{12}$$

$$+ 9\dfrac{5}{6} = + 9\dfrac{10}{12}$$

$$17\dfrac{19}{12} = 18\dfrac{7}{12}$$

4. Add: $4\dfrac{5}{6} + 5\dfrac{7}{8}$

5. Subtract: $24\dfrac{1}{4} - 3\dfrac{4}{5}$

$$24\dfrac{1}{4} = 24\dfrac{5}{20} \qquad 23\dfrac{25}{20}$$

regroup

$$-\ 3\dfrac{4}{5} = 3\dfrac{16}{20} \qquad 3\dfrac{16}{20}$$

$$20\dfrac{9}{20}$$

5. Subtract: $51\dfrac{1}{3} - 26\dfrac{5}{6}$.

6. Estimate: $19\dfrac{5}{6} + 4\dfrac{2}{3} - 6\dfrac{1}{9}$

Estimate $19\dfrac{5}{6}$ rounds up to 20

Estimate $4\dfrac{2}{3}$ rounds up to 5

Estimate $6\dfrac{1}{9}$ rounds down to 6

$20 + 5 - 6 = 19$

$$19\dfrac{120}{144} + 4\dfrac{96}{144} - 6\dfrac{128}{144} = 17\dfrac{88}{144} = 17\dfrac{11}{18}$$

$17\dfrac{11}{18}$ is close to 18

6. Estimate: $28\dfrac{3}{4} + 17\dfrac{1}{6} - 12\dfrac{4}{11}$

Review this example for Objective B:

Solve applied problems involving the addition or subtraction of fractions or mixed numbers.

7. For a family barbeque, Raford bought packages of baby back ribs weighing $2\dfrac{3}{5}$ lb and $5\dfrac{1}{2}$ lb. What was the total weight of the meat?
Add the weights.

$$2\dfrac{3}{5} = 2\dfrac{6}{10}$$

$$+\ 5\dfrac{1}{2} = 5\dfrac{5}{10}$$

$$7\dfrac{11}{10} = 8\dfrac{1}{10}$$

7. Quentin bought $\dfrac{1}{4}$ lb of Swiss cheese, and $\dfrac{2}{5}$ lb of provolone cheese, and $\dfrac{5}{8}$ lb of American cheese. How many pounds of cheese did he buy?

ADDITIONAL EXERCISES

Objective A Add or subtract fractions or mixed numbers.

Perform the indicated operation and simplify.

1. $\dfrac{5}{8} - \dfrac{3}{8}$

2. $\dfrac{8}{12} + \dfrac{7}{20}$

3. $\dfrac{8}{15} - \dfrac{4}{9}$

4. $\dfrac{11}{12} - \dfrac{3}{5}$

5. $9\dfrac{2}{7} - 4\dfrac{4}{7}$

6. $11\dfrac{1}{2} + 5\dfrac{1}{3}$

7. $13\dfrac{1}{3} - 7\dfrac{3}{4}$

8. $7\dfrac{5}{6} - 4$

Estimate each of the following as a whole number.

9. $40\dfrac{7}{9} + 13\dfrac{1}{4}$

10. $56\dfrac{11}{15} - 14\dfrac{9}{10}$

Objective B Solve applied problems involving the addition or subtraction of fractions or mixed numbers.

Solve. Write the answer in simplest form.

11. A house occupies $\dfrac{3}{20}$ acre on a $\dfrac{1}{2}$ acre plot of land. What is the area of the land not occupied by the house?

12. There is $\dfrac{1}{3}$ cup of vegetable oil in a measuring cup. How much oil must be added to make a total of $\dfrac{3}{4}$ cup of oil in the measuring cup?

13. Kevin is $4\dfrac{7}{12}$ inches taller than Marques. Marques is $66\dfrac{3}{4}$ inches tall. How tall is Kevin?

14. Heather uses three pieces of cherry wood to make a drawer front. The middle piece is $11\dfrac{3}{4}$ in. wide, and the two end pieces are each $3\dfrac{1}{2}$ in. wide. How wide is the drawer front?

15. Barry's AAU basketball team won the consolation game in a recent tournament. In this game, Barry scored $\dfrac{4}{9}$ of the team's points, and the second highest scorer scored $\dfrac{1}{4}$ of the team's points. What fraction of the points was scored by the rest of the team?

Chapter 2 FRACTIONS

2.4 Multiplying and Dividing Fractions

Learning Objectives
A   Multiply or divide fractions or mixed numbers.
B   Solve applied problems involving the multiplication or division of fractions or mixed numbers.

MATHEMATICALLY SPEAKING
In exercises 1-4 fill in the blank with the most appropriate term or phrase from the given list.

improper fraction   reciprocal   multiply   simplify   proper fraction   divide reverse

1.   To multiply mixed numbers, change each mixed number to its equivalent

    _____.

2.   To _____ fractions, change the divisor to its reciprocal, and multiply the resulting fractions.

3.   To find the product of the fractions $\frac{2}{3}$ and $\frac{4}{7}$, _____ 2 and 4, and 3 and 7.

4.   The fraction $\frac{5}{9}$ is said to be the _____ of the fraction $\frac{9}{5}$.

EXAMPLES AND PRACTICE

Review this example for Objective A:

Multiply or divide fractions or mixed numbers.

1. Multiply: $\dfrac{5}{7} \cdot \dfrac{3}{4}$

Multiply the numerators and denominators.

$$\dfrac{5}{7} \cdot \dfrac{3}{4} = \dfrac{5 \cdot 3}{7 \cdot 4} = \dfrac{15}{28}$$

Practice:

1. Multiply: $\dfrac{2}{3} \cdot \dfrac{7}{9}$

---

2. Multiply: $\dfrac{2}{3} \cdot \dfrac{7}{8}$

$$\dfrac{2}{3} \cdot \dfrac{7}{8} = \dfrac{\overset{1}{\cancel{2}}}{3} \cdot \dfrac{7}{\underset{4}{\cancel{8}}}$$

$$= \dfrac{7}{12}$$

2. Multiply: $\dfrac{2}{5} \cdot \dfrac{9}{10}$

---

3. Multiply: $3\dfrac{3}{4} \cdot 5\dfrac{1}{3}$ and check the product by estimating.

Write each mixed number as an improper fraction then multiply.

$$3\dfrac{3}{4} \cdot 5\dfrac{1}{3} = \dfrac{15}{4} \cdot \dfrac{16}{3}$$

$$= \dfrac{\overset{5}{\cancel{15}}}{\underset{1}{\cancel{4}}} \cdot \dfrac{\overset{4}{\cancel{16}}}{\underset{1}{\cancel{3}}}$$

$$= 20$$

Check:

$3\dfrac{3}{4} \quad 5\dfrac{1}{3}$

$\downarrow \qquad \downarrow$

$4 \;\cdot\; 5 = 20$

3. Multiply: $4\dfrac{1}{4} \cdot 5\dfrac{1}{3}$ and check the product by estimating.

---

4. Divide: $\dfrac{4}{5} \div \dfrac{3}{4}$

Change the divisor to its reciprocal and multiply.

$$\dfrac{4}{5} \div \dfrac{3}{4} = \dfrac{4}{5} \cdot \dfrac{4}{3}$$

$$= \dfrac{16}{15} = 1\dfrac{1}{15}$$

4. Divide: $\dfrac{5}{6} \div \dfrac{3}{4}$

5. Divide: $6\dfrac{3}{8} \div 2\dfrac{3}{4}$

Change each mixed number to an improper fraction; invert and multiply.

$$6\dfrac{3}{8} \div 2\dfrac{3}{4} = \dfrac{51}{8} \div \dfrac{11}{4}$$

$$= \dfrac{51}{8} \cdot \dfrac{4}{11}$$

$$= \dfrac{51}{\cancel{8}_2} \cdot \dfrac{\cancel{4}^1}{11}$$

$$= \dfrac{51}{22} = 2\dfrac{7}{22}$$

5. Divide: $2\dfrac{5}{16} \div 4\dfrac{3}{4}$

Review this example for Objective B:

Solve applied problems involving the multiplication or division of fractions or mixed numbers.

6. A serving of Good 'n Crunchy cereal is $\dfrac{3}{4}$ cup. How many servings are in a box containing 18 cups of cereal?

Divide 18 by $\dfrac{3}{4}$.

$$18 \div \dfrac{3}{4} = \dfrac{18}{1} \cdot \dfrac{4}{3}$$

$$= \dfrac{72}{3}$$

$$= 24$$

There are 24 servings in the box.

6. A shelf in Clayton's library is 33 in. long. Each volume of his encyclopedia is $1\dfrac{3}{8}$ in. thick. How many volumes of his encyclopedia can he place on the shelf?

Name:

Instructor:

Date:

Section:

ADDITIONAL EXERCISES

Objective A Multiply or divide fractions or mixed numbers.

Perform the indicated operations.

1. $\dfrac{15}{24} \times \dfrac{6}{25}$

2. $12\left(3\dfrac{1}{6}\right)$

3. $2\dfrac{1}{6} \div 3\dfrac{1}{4}$

4. $4\dfrac{2}{3} \div 16$

5. $5\dfrac{2}{3} \cdot 4\dfrac{1}{2}$

6. $3\dfrac{5}{8} \cdot \dfrac{2}{3}$

7. $3\dfrac{5}{8} \div \dfrac{3}{4}$

8. $\dfrac{4}{5} \div 2\dfrac{1}{3}$

9. $8\dfrac{3}{4} \div 5 \times \dfrac{4}{9}$

10. $15\dfrac{3}{5} \div 5\dfrac{3}{5} - 2\dfrac{3}{5}$

Estimate each of the following as a whole number.

11. $7\dfrac{5}{6} \cdot 2\dfrac{1}{9}$

12. $26\dfrac{6}{7} \div 3\dfrac{1}{4}$

Objective B Solve applied problems involving the multiplication or division of fractions or mixed numbers.

Solve. Write the answer in simplest form.

13. Karla receives $60 for working a full day doing inventory at a bookstore. How much will he receive for working $\dfrac{3}{4}$ of the day?

14. The cost of attending Mackinaw State University increased by $4400 in one year. Of this increase, $\frac{5}{8}$ of it is for tuition. How much did tuition increase?

15. A recipe calls for $\frac{3}{5}$ lb of pasta for a salad. How much pasta should be used for $\frac{1}{2}$ of the recipe?

16. A pipe $\frac{7}{8}$ yd long is cut into 3 pieces. How long is each piece?

17. A box of cold cereal says that one serving of the cereal with milk provides 6 grams of protein, which is $\frac{3}{20}$ of the U.S. recommended daily allowance (RDA) for protein. What is the U.S. RDA for protein?

18. The Michigan Department of Transportation (MDOT) is repaving $22\frac{1}{2}$ miles of highway. They have completed $\frac{3}{8}$ of the job. How much of the highway do they still have to pave?

Chapter 3 DECIMALS

3.1 Introduction to Decimals

Learning Objectives
A Read or write decimals.
B Find the fraction equivalent to a decimal.
C Compare decimals.
D Round decimals.
E Solve applied problems involving decimals.

MATHEMATICALLY SPEAKING
In exercises 1-5 fill in the blank with the most appropriate term or phrase from the given list.

right	decreasing	thousandths	less
hundredth	left	increasing	ten
multiple	power	greater	tenth

1. The fractional part of a decimal has as it denominator a _____ of 10.

2. The decimal 0.025 is equivalent to the fraction twenty-five _____.

3. The decimals 0.53, 0.6, and 0.64 are written in _____ order.

4. A decimal place is a place to the _____ of the decimal point.

5. The decimal 184.995 rounded to the nearest _____ is 185.00.

EXAMPLES AND PRACTICE

Review this example for Objective A:	**Practice:**

Read or write decimals.

1. Identify the place that the digit 4 occupies.
   a. 0.134
   b. 92.415
   c. 0.7001487
Solution
   a. thousandths
   b. tenths
   c. hundred-thousandths
	1. Identify the place that the digit 1 occupies.
	   a. 0.134
	   b. 92.415
	   c. 0.7004017

Review this example for Objective B:

Find the fraction equivalent to a decimal.

2. Express 0.125 in fractional form and simplify. The 125 becomes the numerator of the fraction. Since the decimal has three decimal places, we make the denominator of the equivalent fraction 1 followed by three zeros (1000). $$\frac{125}{1000} = \frac{1}{8}$$	**2.** Express 0.625 in fractional form and simplify.
3. Express 1.92 as a mixed number. $$1.92 = 1\frac{92}{100} = 1\frac{23}{25}$$	**3.** Express 2.24 as a mixed number.

4. Express each decimal in words.
   a. 0.214
   b. 3.27
   c. 0.7
 	
   a. two hundred fourteen thousandths
   b. three and twenty-seven hundredths
   c. seven tenths
	4. Express each decimal in words.
	   a. 0.924
	   b. 4.73
	   c. 0.4

5. Write each number in decimal notation.

   a. eight tenths

   b. six and twenty-five thousandths

   a. 0.8

   b. 6.025

---

5. Write each number in decimal notation.

   a. nine tenths

   b. fifteen and twenty-one thousandths

Review this example for Objective C:

Compare decimals.

6. Which is larger, 0.813 or 0.87?

Line up the decimal points.

0.813

0.8**7**

Start comparing the digits in order from left to right. Looking at the hundredths place, we see that 7 > 1.

The larger decimal is 0.87.

---

6. Which is smaller, 0.22$\underset{\cdot}{6}$ or 0.26?

Review this example for Objective D:

Round decimals.
7. Round 76.526 to
    a. the nearest tenth
    b. the nearest ten
    c. the nearest hundredth

   a. 76.$\underline{5}$26
The critical digit 2 is less than 5, so we do not add 1 to the underlined digit. Drop all digits to the right of 5. 76.5

   b. $\underline{7}$6.526 becomes 80
The critical digit is 6 so we add 1 to the underlined digit. Drop all digits to the right of 7.

   c. 76.5$\underline{2}$6 becomes 76.53
The critical digit is 6 so we add 1 to the underlined digit. Drop all digits to the right of 2.

---

7. Round 84.295 to
    a. the nearest tenth
    b. the nearest ten
    c. the nearest hundredth

Review this example for Objective E:	
Solve applied problems involving decimals.	
8. The average annual rain level in Johannesburg, South Africa 28.7 inches. Write this decimal in words.	**8.** In one day, the Dow Jones Industrial Average fell by two hundred seven and sixty-one hundredths points. Write this number in decimal notation.
28.7 in words becomes: twenty-eight and seven tenths	

ADDITIONAL EXERCISES
Objective A Read or write decimals.

Write each decimal in words.

1. 0.629

2. 46.1245

Write each number in decimal notation.

3. Seventeen and nine hundred six ten-thousandths

4. Two hundred forty-three and fifty-two hundredths

Objective B Find the fraction equivalent to a decimal.

Find the equivalent fraction or mixed number, reduced to lowest terms.

5. 0.28

6. 1.16

7. 8.062

8. 23.35

Objective C Compare decimals.

Between each pair of numbers, insert the appropriate sign, <, =, or >, to make a true statement.

9. 0.0506   0.0056

10. 4.89   4.98

11. 8.24   8.42   **12.** 5.62   3.269

Rearrange each group of numbers from smallest to largest.

13. 0.07, 0.0725, 0.0718   **14.** 8.278, 8.27, 8.269

Objective D Round decimals.

Round as indicated.

15. 5.2849 to the nearest hundredth   **16.** 3.951 to the nearest tenth

17. 61.6029 to the nearest hundredth   **18.** 89.10062 to the nearest thousandth

Objective E Solve word problems involving decimals.

Solve.

19. The cost of a gallon of premium gasoline at a particular gas station was $3.769. What is this price to the nearest cent?

20. A quality control inspector measured the diameter of a ball bearing to be 1.41 inches. The actual measurement of the ball bearing should be 1.4 inches. Compare the quality control inspector's measurement with the actual measurement.

21. The average salary of six secretaries was computed to be $29,937.16667. What is the average salary to the nearest hundred?

22. Reed used a home comparison chart to determine how much his Charlotte, NC home would cost him in Boston. After performing the calculation, his calculator read 1713461.538. Round this dollar amount to the nearest thousand.

Chapter 3 DECIMALS

3.2 Adding and Subtracting Decimals

> **Learning Objectives**
> A   Add or subtract decimals.
> B   Solve applied problems involving the addition or subtraction of decimals.

MATHEMATICALLY SPEAKING

In exercises 1-4 fill in the blank with the most appropriate term or phrase from the given list.

sum	decimal points	difference	any number
rightmost digits	zeros		

1. To estimate the _____ between 0.58 and 0.213, subtract 0.2 from 0.6.

2. When adding decimals, rewrite the numbers vertically, lining up the

    _____.

3. Inserting _____ at the right end of a decimal does not change its value.

4. To estimate the _____ of 0.72 and 0.461, add 0.7 and 0.5.

EXAMPLES AND PRACTICE

Review this example for Objective A:	**Practice:**
Add or subtract decimals.	
1. Find the sum: $0.693 + 0.7$ Align the decimal points and add. Insert the decimal point in the sum. $\begin{array}{r} 0.693 \\ + \ 0.700 \\ \hline 1.393 \end{array}$	**1.** Find the sum: $1.5 + 2.009$
2. Subtract: $6.7 - 4.8$ Align the decimal points. Subtract; regrouping when necessary. Insert the decimal point in the difference. $\begin{array}{r} 6.7 \\ - \ 4.8 \quad \text{regroup} \\ \hline 1.9 \end{array}$	**2.** Subtract: $13.62 - 9.98$
3. Add: $2.453 + 1.248$. Then check by estimating. $\begin{array}{rcl} 2.453 & \approx & 2 \\ + \ 1.248 & \approx & + \ 1 \\ \hline 3.701 & & 3 \end{array}$ Our exact sum is fairly close to our estimate.	**3.** Subtract: $18.72 - 9.761$. Then check by estimating.
Review this example for Objective B:	
Solve applied problems involving the addition or subtraction of decimals.	
4. Today's high temperature was 89.4°F. Yesterday's high temperature was 82.7°F. How much warmer is it today than yesterday? Subtract the temperatures.	**4.** Carl filled his truck's gas tank and noted the odometer read 83,142.5. After the next filling, the odometer read 83,446.7. How many miles did Carl travel between fillings?

$$89.4$$
$$\underline{-\ 82.7}\quad \text{regroup as necessary}$$
$$6.7$$

It is $6.7°$ warmer today than yesterday.

ADDITIONAL EXERCISES

Objective A Add or subtract decimals.

Find the sum.

1. $203.28 + 3.39$

2. $3.00482 + 5.3527$

3. $0.41 + 17.2 + 6.9$

4. $2.4 + 1.7 + 0.22$

5. $7.2 + 72 + 720$

6. 0.68 inch $+ 1.1$ inches $+ 1.43$ inches

Find the difference.

7. $6.53 - 2.95$

8. $7.003 - 4.5$

9. $28 - 3.507$

10. $6.703 - 1.5$

11. $\$23.41 - \$15.89$

12. $8.324 - (4.08 - 0.006)$

Estimate the sum or difference by rounding to the nearest tenth.

13. $29.338 + 14.2016$

14. $84.375 - 38.52$

Name:

Instructor:

Date:

Section:

Objective B Solve word problems involving the addition or subtraction of decimals.

Solve. Write your answers in simplest form.

15. Aaron's checking account balance was $23.14 before making deposits of $50 and $38.65. What is Aaron's new checking account balance?

16. The populations of three cities in the U.S. were 8.1 million, 3.8 million, and 2.8 million, respectively. What was the total population of these three cites?

17. The electricity meter on Janet's house read 45,681.2 Kwh at the beginning of the month and 46,107.5 at the end of the month. How many Kwh did Janet use for the month?

Chapter 3 DECIMALS

3.3 Multiplying Decimals

Learning Objectives
A   Multiply decimals.
B   Solve applied problems involving the multiplication of decimals.

MATHEMATICALLY SPEAKING
In exercises 1-4 fill in the blank with the most appropriate term or phrase from the given list.

add	three	factors	first factor
four	multiplication	square	two
division	five		

1. To multiply a decimal by 1,000, move the decimal point _____ places to the right.

2. The product of 0.273 and 8.18 has _____ decimal places.

3. When multiplying decimals, the number of decimal places in the product is equal to the total number of decimal places in the _____.

4. To compute the expression $(6.2)^2 + 9.4$, first _____.

EXAMPLES AND PRACTICE

Review this example for Objective A: | **Practice:**

Multiply decimals.

1. Multiply: 0.86
              × 2.4

First multiply 86 by 24.

```
  0.86
× 2.4
──────
  344
 1720
──────
 2064
```

Count the total number of decimal places in the factors.

```
  0.86    ←2 decimal places
× 2.4     ←1 decimal place
──────
  344
 1720
──────
 2.064    ←3 decimal places in the product
```

1. Multiply: 6.83
              × 9.5

2. Multiply 18.243×0.028 and check the answer by estimating.

Multiply the factors to find the exact product. Then round each factor and multiply.

```
  18.243  ≈       18
× 0.028  ≈ ×   0.03
────────        ──────
0.510804        0.54
```

The exact and estimated product are close.

2. Multiply 37.4×1.82 and check the answer by estimating.

Review this example for Objective B:

Solve applied problems involving the multiplication of decimals.

3. Rebecca pays $24.89 per month for satellite service. How much does she pay altogether in one year?

There are 12 months in a year. Multiply 24.89 by 12.

$$
\begin{array}{r}
24.89 \\
\times \quad 12 \\
\hline
4978 \\
24890 \\
\hline
298.68
\end{array}
$$

Rebecca pays $298.68 per year.

3. Jared purchased 3 lb of grapes at $2.39/lb. How much did he spend on grapes?

ADDITIONAL EXERCISES
Objective A Multiply decimals.

Perform the indicated operation and simplify.

1.
$$
\begin{array}{r}
7.6 \\
\times \quad 9 \\
\hline
\end{array}
$$

2.
$$
\begin{array}{r}
3.8 \\
\times\ 0.7 \\
\hline
\end{array}
$$

3.
$$
\begin{array}{r}
63 \\
\times\ 0.005 \\
\hline
\end{array}
$$

4.
$$
\begin{array}{r}
6.83 \\
\times\quad 9.5 \\
\hline
\end{array}
$$

5.
$$
\begin{array}{r}
0.732 \\
\times\quad 0.8 \\
\hline
\end{array}
$$

6.
$$
\begin{array}{r}
3.67 \\
\times\ 0.05 \\
\hline
\end{array}
$$

7.
$$
\begin{array}{r}
9.28 \\
\times\ 0.031 \\
\hline
\end{array}
$$

8. 10×15.91

9. 384.7×1000

10. 100×0.2153

11. $(4.07)(0.13)$

12. $(0.02)(1.7)(25)$

Estimate each of the following products.

13. 0.653×0.046

14. 563.1×3.15

Objective B Solve applied problems involving the multiplication of decimals.

Solve. Check by estimating.

15. The print shop charges $0.85 for each picture up to 12 pictures and $0.40 for each additional picture. How much would it cost Ben for 20 pictures?

16. Mia is paid $12.46 per hour for the first 40 hours of work. She is paid time and a half (one and one-half times) for any overtime exceeding 40 hours. What is her hourly rate for the time over 40 hours?

17. Getting Fit health club charges $249.95 to join plus an additional monthly fee of $58.95. How much will you pay for the first year of membership?

Chapter 3 DECIMALS

3.4 Dividing Decimals

Learning Objectives
A   Find the decimal equivalent to a fraction.
B   Divide decimals.
C   Solve applied problems involving the division of decimals.

MATHEMATICALLY SPEAKING

In exercises 1-4 fill in the blank with the most appropriate term or phrase from the given list.

quotient	three	divisor	decimal
dividend	right	terminating	four
fraction	product	left	repeating
two			

1. An example of a _____ decimal is 0.54545454.

2. To divide a decimal by 100, move the decimal point _____ places to the left.

3. When dividing a decimal by a whole number, the decimal point in the _____ is placed above the decimal point in the dividend.

4. To change a fraction to the equivalent _____, divide the numerator of the fraction by its denominator.

EXAMPLES AND PRACTICE

Review this example for Objective A: **Practice:**

Find the decimal equivalent to a fraction.

1. Convert $1\frac{4}{5}$ to a decimal.

1. Convert $3\frac{1}{4}$ to a decimal.

Convert the mixed number to an improper fraction.

$$1\frac{4}{5} = \frac{9}{5}$$

Divide the numerator by the denominator.

$$
\begin{array}{r}
1.8 \\
5\overline{)9.0} \\
\underline{5} \\
40 \\
\underline{40} \\
0
\end{array}
$$

Review this example for Objective B:

Divide decimals.

2. What is 5.625 divided by 0.45?

2. What is 38.226 divided by 0.69?

$0.45\overline{)5.625}$ Move the decimal point 2 places right.

$$
\begin{array}{r}
12.5 \\
45\overline{)562.5} \\
\underline{45} \\
112 \\
\underline{90} \\
225 \\
\underline{225} \\
0
\end{array}
$$

You can check by multiplying 12.5 by 45. The result should be 562.5.

3. Divide and check by estimating. Round to the nearest hundredth.

$$\frac{3.622}{0.094}$$

Compute the exact answer.

$$0.\underset{3}{094}\overline{)3.\underset{3}{622}}$$

$$
\begin{array}{r}
38.531 \\
94\overline{)3622.000} \\
\underline{282} \\
802 \\
\underline{752} \\
500 \\
\underline{470} \\
300 \\
\underline{282} \\
180 \\
\underline{94} \\
86
\end{array}
$$

Check by estimating:
3.6 is about 4
0.094 is about 0.1
$4 \div 0.1 = 40$ which is close to our exact answer.

3. Divide and check by estimating. Round to the nearest hundredth.

$$27.3\overline{)621.91}$$

Review this example for Objective C:

Solve applied problems involving the division of decimals.

4. A loan of $9,281.88 is to be paid off in 36 equal monthly payments. How much is each payment?

Divide the loan amount by 36.

$$
\begin{array}{r}
257.83 \\
36\overline{)9281.88}
\end{array}
$$

Each payment is $257.83.

4. If one share of Ford Motor Company stock sold for $7.60, how many shares can be purchased for $3,800?

ADDITIONAL EXERCISES

Objective A Find the decimal equivalent to a fraction.

Change to an equivalent decimal.

1. $\dfrac{3}{8}$

2. $\dfrac{84}{100}$

Objective B Divide decimals.

Divide and check.

3. $8\overline{)78}$

4. $14\overline{)63}$

5. $45\overline{)153}$

6. $\dfrac{58.3}{100}$

7. $\dfrac{54}{0.3}$

8. $1.16 \div 0.8$

9. $13.5 \div 4$

10. $46.2 \div 1.2$

Divide, rounding to the nearest hundredth.

11. $11\overline{)96}$

12. $\dfrac{3.95}{0.7}$

Estimate the quotient.

13. $11.54 \div 0.37$

14. $16.2 \div 0.43 + 12.613$

Objective C Solve word problems involving the division of decimals.

15. Addy's electric bill states that she used 426.3 Kwh over a 29-day period. How many Kwh did she average per day?

16. Owasi bought 8 lbs of rib-eye steak for $37.52. How much did he pay per pound?

17. Carl filled his truck's gas tank and noted the odometer read 83,142.5. At the next filling, the odometer read 83,446.7. It took 15.6 gal to fill the tank the second time. How many miles per gallon did the truck get?

Chapter 4 SOLVING SIMPLE EQUATIONS

4.1 Introduction to Basic Algebra

Learning Objectives
A   Translate phrases to algebraic expressions and vice versa.
B   Evaluate an algebraic expression for a given value of the variable.
C   Solve applied problems involving algebraic expressions.

MATHEMATICALLY SPEAKING

In exercises 1-4 fill in the blank with the most appropriate term or phrase from the given list.

arithmetic   constant   evaluate   translate   variable   algebraic

1.   A(n) _____ expression combines variables, constants, and arithmetic operations.

2.   A(n) _____ is a known number.

3.   A(n) _____ is a letter that represents an unknown number.

4.   To _____ an algebraic expression, replace each variable with the given number, and carry out the computation.

EXAMPLES AND PRACTICE

Review this example for Objective A:

Translate phrases to algebraic expressions and vice versa.

1. Translate each algebraic expression to words.

    a. $x + 6$

    b. $y - 7$

    c. $\dfrac{3}{5}x$

Solution: Sample answers

    a. x plus 6

    b. 7 less than y

    c. $\dfrac{3}{5}$ times x

Practice:

1. Translate each algebraic expression to words.

    a. $9 + y$

    b. $3 - w$

    c. $\dfrac{a}{2}$

Review this example for Objective B:

Evaluate an algebraic expression for a given value of the variable.

2. Evaluate each algebraic expression.
    a. $x + 6$ if $x = 15$ $15 + 6 = 21$

    b. $y - 7$ if $y = 12.54$ $12.54 - 7 = 5.54$

    c. $\dfrac{3}{5}x$ if $x = 25$ $\dfrac{3}{5} \cdot 25 = 15$

2. Evaluate each algebraic expression.
    a. $x + 14$ if $x = 8$

    b. $9 - x$, if $x = 3.6$

    c. $\dfrac{2}{3}y$, if $y = 36$

Review this example for Objective C:

Solve applied problems involving algebraic expressions.

3. When you take out a loan for an automobile, each monthly payment has two parts. One part goes toward the principal and the other part goes toward the interest. If the principal payment is $214.56 and the interest is i, write an algebraic expression for the total payment.

Total payment = $214.56 + i$

3. The school record for points scored is 1421. At the end of the season, Sasha had p points but was short of the record. Write an algebraic expression for the number of points Sasha missed the record by.

ADDITIONAL EXERCISES

Objective A Translate phrases to algebraic expressions and vice versa.

Translate each algebraic expression to two different word phrases.

1. $x + 10$

2. $y - 4$

3. $6 - z$

4. $2p$

5. $\dfrac{1}{2}x$

6. $\dfrac{y}{5}$

Translate each word phrase to an algebraic expression.

7. x decreased by 9

8. 15 more than d

9. Twice p

10. x subtracted from 10

11. the product of y and 6.2

12. the quotient of 17 and n

Objective B Evaluate an algebraic expression for a given value of the variable.

Evaluate each expression.

13. $n \div 4$, if $n = 22.8$

14. $8w$, if $w = 9.4$

15. $t - 4.8$, if $t = 12.1$

16. $d + 48.15$, if $d = 316.47$

17. $\dfrac{5}{9}y$, if $y = \dfrac{3}{10}$

18. $\dfrac{15}{n}$, if $n = 0.25$

Objective C Solve applied problems involving algebraic expressions.

Solve.

19. The value of a certain collectible is doubling every 15 years. If the item is worth w dollars now, write an algebraic expression for the value of the item in 15 years.

20. Whenever you make a payment on a mortgage, part of the payment is used to reduce the principle amount owed and the other part goes toward paying the interest owed. If the principle amount owed is $48,629 and the part of the payment going toward reducing the principle is p, write an algebraic expression for the new principle amount owed.

21. The price of an item you see in the store is the wholesale price plus the markup. If the wholesale price of an item is $43.99 and the markup is m, write an expression for the selling price of the item?

Chapter 4 SOLVING SIMPLE EQUATIONS

4.2 Solving Addition and Subtraction Equations

Learning Objectives
A   Translate sentences involving addition or subtraction to equations.
B   Solve addition or subtraction equations.
C   Solve applied problems involving equations with addition or subtraction.

MATHEMATICALLY SPEAKING

In exercises 1-4 fill in the blank with the most appropriate term or phrase from the given list.

constant	subtract	equation	translates
simplifies	variable	add	sentence
expression			

1.   The equation $x + 2 = 8$ _____ to the sentence "2 more than x is 8."

2.   In the equation $x - 5 = 3$, _____ to each side of the equation in order to isolate the variable.

3.   A solution of an equation is a number that when substituted for the _____ makes the equation a true statement.

4.   A(n) _____ is a mathematical statement that two expressions are equal.

Name: Date:

Instructor: Section:

EXAMPLES AND PRACTICE

Review this example for Objective A: | **Practice:**

Translate sentences involving addition and subtraction to equations.

1. Translate each sentence into an equation.

   a. The sum of x and 12 is equal to 34.

   b. y decreased by 3.2 is 15.

Solution

   a. $x + 12 = 34$

   b. $y - 3.2 = 15$

| **Practice:**

1. Translate each sentence into an equation.

   a. 85 decreased by a number is 59.

   b. y increased by 54 is 102.

Review this example for Objective B:

Solve addition or subtraction equations.

2. Solve and check: $y + 12 = 32$

a.

$$y + 12 = 32$$
$$y + 12 - 12 = 32 - 12 \quad \text{subtract 12 from each side of the equation}$$
$$y + 0 = 20$$
$$y = 20$$

Check:

$$y + 12 = 32$$
$$20 + 12 = 32$$
$$32 = 32$$

b. Solve and check: $n - 3.6 = 1.5$
$$n - 3.6 = 1.5$$
$$n - 3.6 + 3.6 = 1.5 + 3.6$$
$$n + 0 = 5.1$$
$$n = 5.1$$

Check:

$$n - 3.6 = 1.5$$
$$5.1 - 3.6 = 1.5$$
$$1.5 = 1.5$$

2. Solve and check:

a. $y + 16 = 38$

b. $n - 5.8 = 12.3$

Review this example for Objective C:

Solve applied problems involving equations with addition or subtraction.

3. After 3 months of medication and therapy, a patient has gained 12 pounds. If he weighs 168 pounds, what was his original weight?

original weight + weight gain = 168

$$x \quad + \quad 12 \quad = 168$$

$$x + 12 = 168$$

$$x + 12 - 12 = 168 - 12$$

$$x = 156$$

His original weight was 156 pounds.

3. The cost of an airline ticket plus taxes and booking fees of $55.70 is $243.60. What was the cost of the ticket before taxes and fees?

ADDITIONAL EXERCISES
Objective A Translate sentences involving addition or subtraction to equations.

Translate each sentence into an equation.

1. x increased by 4 is 20.

2. The difference between a number and $\dfrac{4}{3}$ is $\dfrac{4}{15}$.

3. The total of a number and 7 yields 36.

4. $5\dfrac{3}{4}$ less than a number is the same as $2\dfrac{3}{8}$.

Objective B Solve addition or subtraction equations.

Solve and check.

5. $x - 8 = 19$

6. $y + 6 = 14$

7. $m - 12 = 8$

8. $d + 14 = 27$

9. $x + 1.2 = 4.7$

10. $r - 5.1 = 7.8$

11. $b - \dfrac{5}{3} = 6$

12. $x + 2\dfrac{3}{5} = 7\dfrac{1}{3}$

Objective C Solve applied problems involving equations with addition or subtraction.

Write an equation and solve.

13. Marques ran for 87 yards in his last game. This is 23 yards less than what he ran in a previous game. How many yards did he the previous game?

14. At 2636 lb, the weight of the 2007 Ford Focus Sedan S is 261 lb less than the weight of the 2007 Nissan Sentra 2.0. What is the weight of the Sentra?

15. The highest grade in a math class was 96. This was 3.4 more than the next highest grade. What was the next highest grade?

Chapter 4 SOLVING SIMPLE EQUATIONS

4.3 Solving Multiplication and Division Equations

Learning Objectives
A   Translate sentences involving multiplication or division to equations.
B   Solve multiplication or division equations.
C   Solve applied problems involving equations with multiplication or division.

MATHEMATICALLY SPEAKING
In exercises 1-4 fill in the blank with the most appropriate term or phrase from the given list.

divide	expression	equation	addition	multiplication
checked	substituting	solved	evaluating	multiply

1.   In the equation $\dfrac{y}{3} = 7$, _____ each side of the equation by 3 in order to isolate the variable.

2.   Check whether a number is a solution to an equation by _____ the number for the variable in the equation.

3.   Division and _____ are opposite operations.

4.   The equal sign separates the two sides of an _____.

EXAMPLES AND PRACTICE

| Review this example for Objective A: | Practice: |

Translate sentences involving multiplication or division to equations.

1. Translate each sentence into an equation.

   a. The product of x and 4 is equal to 0.7

   b. Three-fourths of a number is 12.

Solution

   a. $4x = 0.7$

   b. $\dfrac{3}{4}y = 12$

---

1. Translate each sentence into an equation.

   a. 4.2 divided by a number yields 1.3.

   b. Twelve is equal to one-third of some number.

Review this example for Objective B:

Solve multiplication or division equations.

2. Solve and check: $7x = 56$

   a. $7x = 56$

   $\dfrac{7x}{7} = \dfrac{56}{7}$ Divide each side of the equation by 7.

   $x = 8$

Check:

$7x = 56$

$7(8) = 56$

$56 = 56$

   b. $9 = \dfrac{t}{4}$

   $4 \cdot 9 = 4 \cdot \dfrac{t}{4}$

   $36 = t$

Check:

$9 = \dfrac{t}{4}$

$9 = \dfrac{36}{4}$

$9 = 9$

---

2. Solve and check:

   a. $9x = 81$

   b. $7 = \dfrac{w}{3}$

Review this example for Objective C:

Solve applied problems involving equations with multiplication or division.

3. Each of 6 members of a family will receive an equal amount of the proceeds from a life insurance policy. If each member will receive $45,000, what is the value of the policy?

$$\frac{1}{6}x = 45,000$$

$$6 \cdot \frac{1}{6}x = 6 \cdot 45,000$$

$$x = 270,000$$

The value of the policy is $270,000.

3. Sanu forgot how many CD-R disks he had in stock originally, but knows that the 30 remaining is the original number divided by 4. How many CD-R disks did he have in stock originally?

ADDITIONAL EXERCISES
Objective A Translate sentences involving multiplication or division to equations.

Translate each sentence into an equation.

1. The product of 10 and a number is 65.

2. The quotient of a number and 8 is 13.

3. Twice a number is 38.

4. 5 times an amount is the same as 50.

Objective B Solve multiplication or division equations.

Solve and check.

5. $6x = 42$

6. $15y = 45$

7. $\dfrac{x}{4} = 8$

8. $\dfrac{2}{3}b = 12$

9. $1.6p = 48$

10. $75n = 25$

11. $\dfrac{m}{15} = 8.5$

12. $3x = \dfrac{7}{8}$

Objective C Solve applied problems involving equations with multiplication or division.

Write an equation. Solve and then check.

13. There are 46 cans of regular Pepsi in a cooler. This is twice as many cans as the number of cans of diet Pepsi. How many cans of diet Pepsis are in the cooler?

14. A specialty shop sold 186 sweatshirts to Taylor's graduating class. This is $\dfrac{3}{4}$ of her graduating class. How large is Taylor's graduating class?

15. The total cost of Jamaal's car is $11,109.24. If he is to make equal monthly payments of $308.59, how many payments will he make?

Chapter 5 RATIO AND PROPORTION

5.1 Introduction to Ratios

Learning Objectives
A   Write ratios of like quantities in simplest form.
B   Write rates in simplest form.
C   Solving applied problems involving ratios.

MATHEMATICALLY SPEAKING

In exercises 1-5 fill in the blank with the most appropriate term or phrase from the given list.

weight of the units	numerator	unlike	like	different
quotient	simplest form	fractional form		number of units
denominator	same			

1.   A unit rate is a rate in which the number in the _____ is 1.

2.   Quantities that have the same units are called _____ quantities.

3.   A ratio is a comparison of two quantities expressed as a _____.

4.   A ratio is said to be in _____ when 1 is the only
     common factor of the numerator and denominator.

5.   A rate is a ratio of _____ quantities.

EXAMPLES AND PRACTICE

Review this example for Objective A:	**Practice:**

Write ratios of like quantities in simplest form.

1. Write the ratio 12 to 2 in simplest form.	**1.** Write the ratio 6:18 in simplest form.

$$\frac{12}{2} = \frac{12 \div 2}{2 \div 2} = \frac{6}{1}$$

Review this example for Objective B:

Write rates in simplest form.

2. Simplify : 250 miles to 20 gallons

$$\frac{250 \text{ miles}}{20 \text{ gallons}} = \frac{25 \text{ miles}}{2 \text{ gallons}}$$

2. Simplify: 18 pounds to 15 weeks

3. Write as a unit rate.

$2520 in 8 weeks

$$\frac{\$2520}{8 \text{ weeks}} = \frac{\$2520 \div 8}{8 \text{ weeks} \div 8} = \frac{\$315}{1 \text{ week}}$$

3. Write as a unit rate.
350 copies for $14

Review this example for Objective C:

Solve applied problems involving ratios.

4. Two cities are shown to be 22 inches apart on a map. The distance between these cities is 440 miles. How many miles per inch is used for the scale of the map?

$$\frac{440 \text{ miles}}{22 \text{ inches}} = \frac{440 \text{ miles} \div 22}{22 \text{ inches} \div 22} = \frac{20 \text{ miles}}{1 \text{ inch}}$$

4. In 6 months, Robert drove his new car 3,438 miles. How many miles are driven per month?

ADDITIONAL EXERCISES

Objective A Write ratios of like quantities in simplest form.

Write each ratio as a fraction in simplest form.

1. 24 to 30

2. 45 to 27

3. 1.75 to 100

4. 25 to $2\frac{3}{4}$

5. 12 ounces to 40 ounces

6. 48 months to 48 months

Objective B Write rates in simplest form.

Write each rate in simplest form.

7. 24 feet in 15 seconds

8. 245 miles in 10 hours

9. 716 yards in 6 games

10. 8 teachers for 76 students

11. 4 cans of paint for 1,576 square feet of wall

12. $1400 for 5 people

Determine the unit rate.

13. 270 km in 6 hours

14. 340 miles using 17 gallons of gas

15. $620 for 40 hours of work

16. 45 ounces of sugar for 25 servings

Find the unit price.

17. 12 bars of soap for $9.36

18. 16 slices of cheese for $3.99

Objective C Solving applied problems involving ratios.

Solve.

19. Tahir made a 34 minute cell phone call that cost him $5.10. What is the cost per minute of this call?

20. A statistics teacher noticed that 520 of her 780 students over the years have been female. What is the ratio of females to males that have taken her class?

21. On a particular day, 1,000 U.S. dollars were worth 1,232 Euros. What was the exchange rate of Euros to the U.S. dollar?

Chapter 5 RATIO AND PROPORTION

5.2 Solving Proportions

Learning Objectives
A   Write or solve proportions.
B   Solve applied problems involving proportions.

MATHEMATICALLY SPEAKING
In exercises 1-4 fill in the blank with the most appropriate term or phrase from the given list.

    equation   check   like   products   solve   cross products   as   proportion

1.   The proportion $\dfrac{2}{3} = \dfrac{6}{9}$ can be read "2 is to 3 _____ 6 is to 9."

2.   A(n) _____ is a statement that two ratios are equal.

3.   To _____ the proportion $\dfrac{x}{5} = \dfrac{7}{10}$, find the value of x that makes the proportion true.

4.   To determine if a proportion is true, check whether the _____ of the ratios is/are equal.

EXAMPLES AND PRACTICE

| **Review this example for Objective A:** | **Practice:** |

Write or solve proportions.

1. Solve and check: $\dfrac{x}{18} = \dfrac{2}{3}$

1. Solve and check: $\dfrac{13}{15} = \dfrac{w}{75}$

Solution   Check:

$$\dfrac{x}{18} = \dfrac{2}{3}$$

$$3 \cdot x = 18 \cdot 2$$

$$3x = 36$$

$$\dfrac{3x}{3} = \dfrac{36}{3}$$

$$x = 12$$

Check:

$$\dfrac{x}{18} = \dfrac{2}{3}$$

$$\dfrac{12}{18} = \dfrac{2}{3}$$

$$12 \cdot 3 = 18 \cdot 2$$

$$36 = 36$$

Review this example for Objective B:

Solve applied problems involving proportions.

2. For a school picnic, 2 pounds hamburger meat serves 9 people. How many pounds of hamburger would be needed to serve 156 people?

$$\dfrac{2}{9} = \dfrac{x}{156}$$

$$9 \cdot x = 2 \cdot 156$$

$$9x = 312$$

$$\dfrac{9x}{9} = \dfrac{312}{9}$$

$$x = 34\dfrac{2}{3}$$

$34\dfrac{2}{3}$ pounds of hamburger would be needed.

2. An automobile goes 229.5 miles on 9 gallons of gasoline. At the same rate, how far would the automobile go on 20 gallons of gasoline?

Objective A Write or solve proportions.

Solve and check.

1. $\dfrac{x}{8} = \dfrac{9}{6}$

2. $\dfrac{6}{11} = \dfrac{12}{n}$

3. $\dfrac{7}{4} = \dfrac{x}{8}$

4. $\dfrac{24}{x} = \dfrac{16}{12}$

5. $\dfrac{68}{y} = \dfrac{17}{25}$

6. $\dfrac{x}{5} = \dfrac{5}{2}$

7. $\dfrac{7}{3} = \dfrac{2}{x}$

8. $\dfrac{x}{13} = \dfrac{2}{9}$

9. $\dfrac{9}{1\frac{1}{4}} = \dfrac{x}{3}$

10. $\dfrac{30}{6.4} = \dfrac{x}{8}$

Objective B Solve applied problems involving proportions.

Solve and check.

11. A car is driven 37,500 miles in 3 years. At that rate, in how many years will the car have 100,000 miles on it?

12. Based on a recent study, 6 out of every 10 Americans are considered overweight. Currently there are about 300 million Americans. How many would be considered overweight?

13. Lauren studied 12 hours and got a score of 80 on her exam. At this rate, how long would she have to study in order to get a score of 95?

14. A family travels 1,230 miles in 4 days. At the same rate, how long would the family take to travel 5,535 miles?

15. Sylvia used 2 gallons of paint to paint 390 ft^2. She has 2,340 ft^2 remaining to paint. How many gallons of paint will she need in order to complete the job?

16. On a 50-pound bag of fertilizer, the directions say to apply 2.5 lb of fertilizer for every 100 square feet. How many pounds of fertilizer are needed for 2,500 square feet?

17. The Trouts pay $2,670 in property taxes on their house valued at $178,000. At this rate, how much would the McDonalds pay in taxes on their $210,000 home?

Chapter 6 PERCENTS

6.1 Introduction to Percents

Learning Objectives
A   Find the fraction or the decimal equivalent of a given percent.
B   Find the percent equivalent of a given fraction or decimal.
C   Solve applied problems involving percent conversions.

MATHEMATICALLY SPEAKING
In exercises 1-4 fill in the blank with the most appropriate term or phrase from the given list.

right   **fraction**   **percent**
decimal   **left**

1. To change a percent to the equivalent decimal, move the decimal point two places to the
    _____ and drop the percent sign.

2. To change a fraction to the equivalent percent, change the fraction to a
    _____, which in turn is changed to a percent.

3. A _____ is a ratio or fraction with denominator 100.

4. To change a percent to the equivalent _____, drop the % sign from
    the given percent, and place the number over 100.

EXAMPLES AND PRACTICE

Review this example for Objective A:	**Practice:**

Find the fraction or the decimal equivalent of a given percent.

1. a. Write 8% as a fraction.

Drop the percent sign and write the 8 over 100 and reduce.

$$\frac{8}{100} = \frac{8 \div 4}{100 \div 4} = \frac{2}{25}$$

b. Write 250% as a fraction.

$$\frac{250}{100} = \frac{5}{2} \text{ or } 2\frac{1}{2}$$

1. a. Write 12% as a fraction.

b. Write 350% as a fraction.

---

2. a. Change 45% to a decimal.

Drop the % sign and divide by 100; or move the decimal point two places to the left.

45% = 0.45

b. Find the decimal equivalent of 2%.

2% = 0.02

2. a. Change 55% to a decimal.

b. Change 5% to a decimal.

Review this example for Objective B:

Find the percent equivalent of a given fraction or decimal.

3. a. Write 0.379 as a percent.

Multiply 0.379 by 100 and add a percent sign.
$0.379 \times 100\% = 37.9\%$

b. Convert 0.04 to a percent.

Move the decimal point two places to the right.
$0.04 = 004.\% = 4\%$

3. a. Write 0.254 as a percent.

b. Convert 0.12 to a percent.

4. Write $\dfrac{9}{20}$ as a percent.

To change the given fraction to a percent, multiply by 100 and insert a % sign.

$$\dfrac{9}{20} = \dfrac{9}{20} \times 100\% = \dfrac{9}{\overset{}{\underset{1}{\cancel{20}}}} \times \dfrac{\overset{5}{\cancel{100}}\%}{1} = 45\%$$

4. Write $\dfrac{6}{25}$ as a percent.

Review this example for Objective C:

Solve applied problems involving percent conversions.

5. At one point the U.S. national unemployment rate was 5.5%. Express this percent as a decimal.

$5.5\% = 0.055$

5. A recent ad for pain medicine indicated that 7 out of 10 doctors preferred a certain brand. What percent of doctors preferred this brand?

ADDITIONAL EXERCISES

Objective A Find the fraction or the decimal equivalent of a given percent.

Change each percent to a fraction or mixed number, and simplify.

1. 12%

2. 63%

3. 120%

4. $\frac{4}{5}\%$

5. $\frac{1}{8}\%$

6. $8\frac{1}{4}\%$

Change each percent to a decimal.

7. 46%

8. 175%

9. 72.5%

10. 200%

11. $9\frac{3}{4}\%$

12. $114\frac{7}{10}\%$

Objective B Find the percent equivalent of a given fraction or decimal.

Express each decimal as a percent.

13. 0.08

14. 0.375

15. 1.72

16. 0.0025

Change each fraction to a percent.

17. $\frac{3}{4}$

18. $\frac{7}{25}$

19. $\dfrac{4}{9}$ **20.** $\dfrac{13}{50}$

Objective C Solve applied problems involving percent conversions.

Solve.

21. In Chicago, 27.6% of all workers take public transit to work. Express this percent as a decimal.

22. The value of a car typically decreases by 30% in the first year. Express this percent as a fraction.

23. On a recent test, Maria got $\dfrac{4}{5}$ of the questions correct. What percent of the questions did she get correct?

24. The population of Novi, Michigan increased 110% from 1980 to 2000. Express this percent as a simplified mixed number.

Chapter 6 PERCENTS

6.2 Solving Percent Problems

Learning Objectives
A   Find the amount, the base, or the percent in a percent problem. B   Solve applied problems involving percents.

MATHEMATICALLY SPEAKING

In exercises 1-4 fill in the blank with the most appropriate term or phrase from the given list.

amount	of	base	is
what	percent		

1. In the translation method of solving a percent problem, _____ is replaced by a multiplication symbol.

2. The _____ of the base is the amount.

3. The _____ is the result of taking the percent of the base.

4. The _____ is the number that we are taking the percent of.

EXAMPLES AND PRACTICE

Review this example for Objective A:

Find the amount, the base, or the percent in a percent problem.

1. **a.** What is 20% of 70?

$$x = 0.20 \cdot 70$$

$$x = 14$$

14 is 20% of 70.

b. 48 is what percent of 124?

$$48 = x \cdot 124$$

$$48 = 124x$$

$$\frac{48}{124} = \frac{124x}{124}$$

$$0.387 = x$$

48 is about 39% of 120.

c. 25% of what number is 40?

$$0.25 \cdot x = 40$$

$$0.25x = 40$$

$$\frac{0.25x}{0.25} = \frac{40}{0.25}$$

$$x = 160$$

25% of 160 is 40.

Practice:

1. **a.** What is 30% of 80?

b. 42 is what percent of 110?

c. 25% of what number is 60?

Review this example for Objective B:

Solve applied problems involving percents.

2. The value of a car typically decreases by 30% in the first year. A car is purchased for $16,000. What is the value of the car one year after it was purchased?

What is 30% of 16, 000?

$x = 0.30 \cdot 16{,}000$

$x = 4800$

The value of the car is
$\$16{,}000 - \$4{,}800 = \$11{,}200$

2. Austin has 4% of his salary taken out for his 401(k). If his annual salary was $54,600, how much did he put into his 401(k)?

ADDITIONAL EXERCISES

Objective A Find the amount, the base, or the percent in a percent problem.

Find the amount.

1. What is 12% of 150.

2. 125% of 420 is what?

3. What is 2.7% of $8,000?

4. $\frac{1}{5}$% of 70 is what?

Find the base.

5. 18 is 60% of what number?

6. $42 is 28% of what amount of money?

7. 8 is 300% of what number.

8. 12 miles is 30% of what length?

Find the percent.

9. What percent of 30 is 4.5?

10. What percent of 16 is 28?

11. $475 is what percent of $1,900?

12. 68 is what percent of 80?

Objective B Solve applied problems involving a percent.

Solve.

13. On a test of 40 items, Kahli got 32 correct. What percent did Kahli get correct?

14. Corey spent 6 hours playing video games in one day. If he was awake for 15 hours, what percent of his waking hours did he spend playing video games?

15. The Grizzlies won 65% of their games last year. If they won 26 games, how many games did they play?

Chapter 6 PERCENTS

6.3 More on Percents

Learning Objectives
A   Solve percent increase or decrease problems.
B   Solve percent problems involving taxes, commissions, and discounts.
C   Solve simple or compound interest problems.
D   Solve applied problems involving percents.

MATHEMATICALLY SPEAKING

In exercises 1-4 fill in the blank with the most appropriate term or phrase from the given list.

discount	salary	commission
markup	simple	compound
final	original	

1.   When computing a percent increase or decrease, the _____ value is used as the base of the percent.

2.   _____ interest is paid on both the principal and the previous interest generated.

3.   Sellers who are paid a fixed percent of the sales for which they are responsible are said to work on _____.

4.   A reduction on the price of merchandise is called a _____.

EXAMPLES AND PRACTICE

Review this example for Objective A:

Solve percent increase or decrease problems.

1. A puppy weighed 3.5 pounds at birth and now weighs 5.6 pounds. What is the percent increase?

Find the difference between the two values.

$5.6 - 3.5 = 2.1$

What percent of 3.5 is 2.1?

$x \cdot 3.5 = 2.1$

$3.5x = 2.1$

$\dfrac{3.5x}{3.5} = \dfrac{2.1}{3.5}$

$x = 0.6$

The percent increase is 60%.

Review this example for Objective B:

Solve percent problems involving taxes, commissions, and discounts.

2. A coffee table marked $299 is on sale for $227. Find the discount and the discount rate, to the nearest percent.

The discount is $299 – $227 = $72

$72 is what percent of $299?

$72 = 299x$

$\dfrac{72}{299} = \dfrac{299x}{299}$

$0.2408 \approx x$

The discount rate is about 24%.

Practice:

1. Enrollment in a mathematics course went from 62 students last semester to 58 students this semester. What is the percent decrease? Round to the nearest whole.

2. A purchase of $44.00 has a sales tax of $2.31. What is the sales tax rate?

3. A real estate agent receives a 6% commission on the sale of a $225,000 home. What was her commission?

What is 6% of 225,000?

x = 0.06 · 225,000

x = 13,500

Her commission was $13,500.

3. A real estate agent receives a 5% commission on the sale of a $190,000 home. What was her commission?

Review this example for Objective C:

Solve simple or compound interest problems.

4. Taylor invests $800 in a 9-month CD that pays 2% simple interest. What is the value of the investment at the end of the 9-month period?

Interest = Principal × Rate × Time

= 800(0.02)(9/12)

= 12

The investment earned $12.

The value of the investment is:
Original deposit + Interest

$800 + $12 = $812

4. Trina invests $700 in a 9-month CD that pays 3% simple interest. What is the value of the investment at the end of the 9-month period?

---

5. Find the amount in an account if $7,500 is invested at 3.5%, compounded annually, for 2 years.

Interest = Principal × Rate × Time

= 7,500(0.035)(1)

= 262.50

Amount in account after first year.
$7,500 + $262.50 = $7,762.50

Find the interest the second year.
Interest = Principal × Rate × Time

= 7,762.50(0.035)(1)

= 271.69

Amount in account after year 2.
$7,762.50 + $271.69 = $8,034.19

5. Find the amount in an account if $6,500 is invested at 2.5%, compounded annually, for 2 years.

ADDITIONAL EXERCISES
Objective A Solve percent increase or decrease problems.

Solve.

1. Tuition at a local college increased from $6,890 to $8,215. What was the percent increase?

2. During the down national economy, home values dropped. A house valued at $275,000 was now valued at $255,750. What was the percent decrease?

3. As a result of dieting, Sonata goes from a weight of 140 lb to 128 lb. What is the percent decease?

4. Reed's salary was $36,455 in 2011. This was an increase of $4,755 since 2007. What was the percent increase?

5. The population of the Charlotte, N.C. metropolitan area was 1,475,000 in 2004. This was a 450,000 increase since 2000. What was the percent increase, to the nearest percent?

Objective B Solve percent problems involving taxes, commissions, and discounts.

Solve.

6. A real estate agent receives a 7% commission on the sale of a $155,000 home. What was her commission?

7. Regal, Inc. pays all of their salespeople a monthly base pay of $2,900 and an additional 5% commission on their monthly sales. Find the monthly salary of a salesperson having sales of $32,780 for the month.

8. The sales tax rate in Michigan is 6%. Compute the sales tax on a large-screen TV priced at $1,948.

9. A department store advertises that you get an additional 15% discount on items purchased with the store credit card. If your total purchase is $417.38, what amount will you actually pay if you use your store credit card?

10. An electronics store sells a digital camcorder for $569.99. This price is a 35% markup of the price the store paid for the camcorder. How much did the store pay for the camcorder?

11. A discontinued printer that regularly sells for $175 is on clearance for 30% off. The tax rate is 7.5%. Find the discount, the tax, and the total cost of the printer.

Objective C Solve simple or compound interest problems.

Solve.

12. What is the interest on $3,400 invested at an interest rate of 8% for 1 year?

13. What is the interest on $1,800 invested at an interest rate of 7% for $\frac{1}{2}$ year?

14. Sanchez puts $2,600 in an account with an annual interest rate of 6%. What is the account balance after 3 years (simple interest)?

15. If $5,000 is invested in an account that pays 6.5% annual interest, what is the account balance after 2 years(simple interest)?

16. Find the amount in an account if $4,000 is invested at 5%, compounded annually, for 2 years.

Objective D was covered in the previous three objectives.

Chapter 7 SIGNED NUMBERS

7.1 Introduction to Signed Numbers

Learning Objectives
A   Represent signed numbers on a number line.
B   Find the opposite of a signed number.
C   Find the absolute value of a signed number.
D   Compare signed numbers.
E   Solve applied problems involving signed numbers.

MATHEMATICALLY SPEAKING

In exercises 1-5 fill in the blank with the most appropriate term or phrase from the given list.

larger	smaller	opposites	positive number
integers	absolute value	signed number	negative number
whole number			

1.   A number less than 0 is a _____*negative number*_____.

2.   The _____*integers*_____ are the numbers $\cdots, -4, -3, -2, -1, 0, 1, 2, 3, 4, \cdots$, counting indefinitely in both directions.

3.   The _____*absolute value*_____ of a number is the distance from 0 on the number line.

4.   For two numbers on the number line, the number on the right is _____*larger*_____ than the number on the left.

5.   Two numbers that are the same distance from 0 on the number line but on opposite sides of 0 are called _____*opposites*_____.

Name: Maxyn Hallare   Date: 4/8/2016

Instructor: Section:

EXAMPLES AND PRACTICE

Review this example for Objective A:

Represent signed numbers on a number line.

1. Locate $\frac{1}{4}$, -3.5, $-\frac{3}{4}$ and 1.8 on the number line.

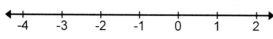

Solution

-3.5 $-\frac{3}{4}$ $\frac{1}{4}$ 1.8

Practice:

1. Locate $\frac{1}{2}$, -1.4, $-2\frac{3}{4}$ and 1.27 on the number line.

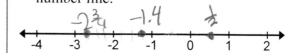

Review this example for Objective B:

Find the opposite of a signed number.

2. Find the opposite of each number.
   a. 9
   b. -112

Two numbers are opposites if they are the same distance from 0 on the number line but on opposite sides of 0.

Solution
a. The opposite of 9 is -9.
b. The opposite of -112 is 112.

2. Find the opposite of each number.
   a. 15
   b. -87

a. -15
b. 87

Review this example for Objective C:

Find the absolute value of a signed number.

3. Determine the sign and the absolute value of the number.
   a. 36
   b. -2.3

   a. Sign: +; absolute value: 36
   b. Sign: −; absolute value: 2.3

3. Determine the sign and the absolute value of the number.
   a. 21
   b. -3.9

a. +, 21
b. −, 3.9

Review this example for Objective D:	
Compare signed numbers. **4.** Which is larger –2 or –4?	**4.** Which is larger –1 or –1.6?

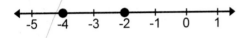

Because –2 is to the right of –4, –2 > –4; that is, –2 is larger.

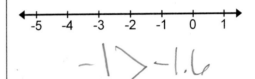

$-1 > -1.6$

Review this example for Objective E:

Solve applied problems involving signed numbers.

5. A submarine descends one hundred twenty feet. Express this quantity as a signed number.

Solution: –120 feet

5. Would the balance in a checking account be greater or less if a check for $50 or a check for $30 was written?

greater if check for $30 was written

ADDITIONAL EXERCISES

Objective A Represent signed numbers on a number line.

Locate each number on the number line.

1. $-2, 3, 1\frac{4}{5}$

Objective B Find the opposite of a signed number.

Find the opposite of each number.

2. $14 \rightarrow -14$

3. $-3\frac{5}{9} \rightarrow 3\frac{5}{9}$

Objective C Find the absolute value of a signed number.

Evaluate.

4. $|-16|$ 16

5. $|-3.2|$ 3.2

6. $\left|\dfrac{9}{20}\right|$ $\dfrac{9}{20}$

7. $\left|-2\dfrac{1}{6}\right|$ $2\dfrac{1}{6}$

Objective D Compare signed numbers.

Indicate the larger of the two numbers.

8. -5 and -2

9. 4 and -8

10. -3.4 and -3.6

11. $-4\dfrac{2}{5}$ and $-4\dfrac{3}{5}$

Objective E Solve applied problems involving the comparison of signed numbers.

Indicate the larger of the two numbers.

12. Marge's checking account statement showed a balance of –$14.89. Randy's checking account statement showed a balance of –$23.37. Whose account was most overdrawn?

13. While playing "Jeopardy," contestants Mack and Stacey both have the same score (dollar amount). In successive rounds, Mack misses an $800 question and then a $100 question. Also, Stacey misses two $400 questions. Which contestant now has the highest score?

14. In January, the temperature was 6° below zero on a Tuesday and 9° below zero on Wednesday. Which day had the highest temperature?

15. Would an airplane be higher if it descended 1200 ft or 1400 feet?

Chapter 7 SIGNED NUMBERS

7.2 Adding Signed Numbers

Learning Objectives
A   Add signed numbers.
B   Solve applied problems involving the addition of signed numbers.

MATHEMATICALLY SPEAKING
In exercises 1-4 fill in the blank with the most appropriate term or phrase from the given list.

commutative	**right**	**larger**	**absolute value**
left	**distributive**	**smaller**	**numbers**

1.   To add (–8) and (+8) on the number line, start at (–8) and move 8 units to the

_____.

2.   Rearranging signed numbers to add the positives and negatives separately does not
     affect the sum because the operation of addition is associative and

_____.

3.   To find the sum of two signed numbers with different signs, subtract the smaller absolute
     value from the larger and take the sign of the number with the _____
     absolute vale.

4.   To find the sum of two signed numbers with the same sign, add the
     _____ and keep the sign.

EXAMPLES AND PRACTICE

Review this example for Objective A:	**Practice:**
Add signed numbers.	
1. Add: −6 and −2	**1.** Add: −7 and −3
The sum of the absolute values is 8.	
The numbers are both negative, so their sum is negative.	
$(-6) + (-2) = -8$	
2. Add: $(-2) + 7$	**2.** Add: $(-3) + 9$
We are adding numbers with different signs. Find the absolute values.	
$\lvert -2 \rvert = 2 \qquad \lvert 7 \rvert = 7$	
Subtract the smaller absolute value from the larger. $7 - 2 = 5$	
Because 7 has the larger absolute value and its sign is positive, the sum is also positive.	
$(-2) + 7 = 5$	
Review this example for Objective B:	
Solve applied problems involving the addition of signed numbers.	
3. One day the temperature in Butte, Montana was 6°F below zero. If the temperature rose by 15° by 6:00, what was the temperature at 6:00?	**3.** One day the temperature in Butte, Montana was 8°F below zero. If the temperature rose by 12° by 4:00, what was the temperature at 4:00?
$-6 + 15 = 9°F$	

ADDITIONAL EXERCISES
Objective A Add signed numbers.

Find the sum.

1. $-4+(-5)$

2. $16+(-9)$

3. $-11+24$

4. $-18+(-15)$

5. $-8.7+15.2$

6. $15.6+(-19.2)$

7. $-\dfrac{3}{4}+\left(-\dfrac{2}{5}\right)$

8. $-\dfrac{1}{4}+\dfrac{7}{9}$

9. $1\dfrac{4}{5}+\left(-3\dfrac{1}{2}\right)$

10. $-5\dfrac{3}{8}+\left(-2\dfrac{1}{4}\right)$

Objective B Solve applied problems involving signed numbers.

Solve. Express each answer as a signed number.

11. One day, the temperature in Des Moines, Iowa was 30°F at noon. A cold front moved in and temperature dropped 42° by midnight. What was the temperature at midnight?

12. United Motors offered a buyout program for employees, and 279 employees enrolled. One year later, another buyout plan was offered, and 131 employees enrolled. What is the change in the number of employees working for the company as a result of the two buyout programs?

13. Power Division, Inc. posted a profit of $10,500 in 2009. In 2010, they posted a $16,600 loss. What is the company's total profit over the two year period?

14. Boris has $312 in a checking account. He writes a check for $346. What is the balance in his account?

15. The base of the tallest mountain in the world, Mauna Kea in Hawaii, is 19,684 ft below sea level. Its peak is 33,480 ft above its base. What is the elevation of the peak above sea level?

Chapter 7 SIGNED NUMBERS

7.3 Subtracting Signed Numbers

Learning Objectives
A   Subtract signed numbers.
B   Solve applied problems involving the subtraction of signed numbers.

MATHEMATICALLY SPEAKING

In exercises 1-4 fill in the blank with the most appropriate term or phrase from the given list.

absolute value	**order of operations**	**addition**
sum	**multiplication**	**signed numbers**
difference	**opposite**	

1. When a signed number problem involves addition and subtraction, work from left to right according to the _____ rule.

2. Every subtraction problem has a related _____ problem.

3. To subtract two signed numbers, change the operation of subtraction to addition, and change the number being subtracted to its _____. Then follow the rule for adding signed numbers.

4. When subtracting a negative number, the _____ is greater than the original number.

Name: Date:
Instructor: Section:

EXAMPLES AND PRACTICE

Review this example for Objective A:	Practice:

Subtract signed numbers.

1. Find the difference: $-3 - (-5)$

We change the operation of subtraction to addition. Then we add the first number and the opposite of the second number.

$-3 - (-5)$
$-3 + (+5) = 2$

1. Find the difference: $-4 - (-9)$

---

2. Compute: $8 - (-6)$

Change the subtraction sign to addition and the -6 to its opposite 6.

$8 - (-6) = 8 + 6 = 14$

2. Compute: $9 - (-7)$

---

3. Subtract: $-4 - 5\dfrac{1}{8}$

$-4 - 5\dfrac{1}{8} = -4 + \left(-5\dfrac{1}{8}\right) = -9\dfrac{1}{8}$

3. Subtract: $-2 - 2\dfrac{1}{5}$

---

4. Calculate: $2 + (-4) - (-12)$
This problem involves addition and subtraction. According to the order of operations rule, we work from left to right.
$2 + (-4) - (-12) = -2 - (-12)$ Add 2 and (-4)
$\qquad\qquad\qquad = -2 + 12$ Subtract -12
$\qquad\qquad\qquad = 10$

4. Calculate: $7 + (-8) - (-14)$

Review this example for Objective B:

Solve applied problems involving the subtraction of signed numbers.

5. The boiling point of acetylene is $-119.2°$ F. The temperature at which butylene boils is $21.2°$ F. What is the difference between these two temperatures?

$$21.2 - (-119.2) = 21.2 + 119.2$$
$$= 140.4$$

The difference in temperatures is $140.4°$ F.

5. The boiling point of methane is $-258.7°$ F. The temperature at which triptane boils is $177.6°$ F. What is the difference between these two temperatures?

ADDITIONAL EXERCISES
Objective A Subtract signed numbers.

Find the difference.

1. $-15 - 9$

2. $-48 - (-15)$

3. $12 - (-8)$

4. $-7 - (-10.8)$

5. $87 - 29$

6. $15.6 - 19.2$

7. $-\dfrac{1}{4} - \left(-\dfrac{3}{5}\right)$

8. $-\dfrac{3}{4} - \dfrac{8}{9}$

9. $2\dfrac{2}{5} - \left(-5\dfrac{3}{8}\right)$

10. $-12 - \left(-4\dfrac{3}{7}\right)$

Combine.

11. $18-(-15)-3+2$

12. $-31+(-28)-(-14)$

13. $14.9-(-50.7)+20-(-32.8)$

14. Subtract -16 from -7.

Objective B Solve applied problems involving the subtraction of signed numbers.

Solve. Express each answer as a signed number.

15. In 1947, Chuck Yeager flew the Bell X-1 "Glamorous Glennis" at 70,104 ft, the highest altitude reached by a manned aircraft up to that time. In 1962, Joseph Walker reached an altitude of 354,300 ft flying the X-15. How much higher did Joseph Walker fly than Chuck Yeager?

16. The elevation of the highest point in Africa is Mt. Kilimanjaro, Tanzania at 19,340 ft. The lowest elevation is at Lake Assal, Djibouti, 512 ft below sea level. What is the difference in the elevations of the two locations?

17. Preston has a balance of $519.43 on his credit card. He returns a sweater that cost him $84.79. How much does he now owe on his credit card?

18. In one of the closest 100 meter finishes ever, Veronica Campbell of Jamaica won in 11.01 seconds at the 2007 World Championships in Osaka, Japan. Christine Arron of France finished sixth in 11.08 seconds. What is the difference between the winning time and the sixth place time?

Chapter 7 SIGNED NUMBERS

7.4 Multiplying Signed Numbers

Learning Objectives
A   Multiply signed numbers.
B   Solve applied problems involving the multiplication of signed numbers.

MATHEMATICALLY SPEAKING

In exercises 1-4 fill in the blank with the most appropriate term or phrase from the given list.

odd   **positive**   **negative**   **even**

1. The product of a(n) _____ number of negative factors is positive.

2. The product of two numbers with the same sign is _____.

3. The product of a(n) _____ number of negative factors is negative.

4. The product of two numbers with different signs is _____.

EXAMPLES AND PRACTICE

Review this example for Objective A:	Practice:
Multiply signed numbers.	
1. Calculate: $(8)(-7)$ Find the absolute value of the factors $|8| = 8$ and $|-7| = 7$ $8 \cdot 7 = 56$ The factors have different signs, so the product is negative. $(8)(-7) = -56$	**1.** Calculate: $(9)(-4)$
2. Multiply: $(-1.2)(-0.5)$ Multiply the absolute values. $(1.2)(0.5) = 0.6$ Since the factors have the same sign, we get $(-1.2)(-0.5) = 0.6$	**2.** Multiply: $(-1.4)(-0.6)$
Review this example for Objective B: **Solve applied problems involving the multiplication of signed numbers.** **3.** Galaxy Motors posted a net loss of $320,000 in the 2010. If this trend continues for the next 4 years, how much more money would Galaxy Motors lose? $-320,000(4) = -\$1,280,000$	**3.** Honey Bear Ice Cream posted a net loss of $215,000 in the 2010. If this trend continues for the next 3 years, how much more money would Honey Bear Ice Cream lose?

ADDITIONAL EXERCISES
Objective A Multiply signed numbers.

Find the product.

1. $9 \cdot (-8)$ **2.** $-21 \cdot 5$

3. $-8 \cdot (-13)$

4. $(-300) \cdot (-2)$

5. $-30 \cdot 60$

6. $(0.7) \cdot (0.6)$

7. $(-5)(-1.3)$

8. $\dfrac{3}{4} \cdot \left(-\dfrac{16}{27} \right)$

Evaluate.

9. -6^2

10. $(-0.3)^2$

Compute.

11. $2(-5)(-75)$

12. $(4)(-2)(3)(-1)$

Objective B Solve applied problems involving the multiplication of signed numbers.

Solve. Express each answer as a signed number.

13. The value of a copier is decreasing by $300 per year. At this rate, how much would the copier lose in value in 4 years?

14. A home builder loses $1,840 per month for a house that does not sell since the completion of its building. How much will the home builder lose if a house was unsold for 8 months?

15. Galena's cell phone bill is set at $55 per month. How much did she spend on cell phone use for the year?

16. The temperature in Chicago dropped an average of 2°F per hour for 6 hours. If the initial temperature was 9°, what was the temperature at the end of the 6 hour period?

Chapter 7 SIGNED NUMBERS

7.5 Dividing Signed Numbers

Learning Objectives
A   Divide signed numbers.
B   Solve applied problems involving the division of signed numbers.

MATHEMATICALLY SPEAKING

In exercises 1-4 fill in the blank with the most appropriate term or phrase from the given list.

addition   positive   negative   unequal   equal   multiplication

1. Every division problem has a related _____ problem.

2. The fractions $\dfrac{-7}{10}$, $\dfrac{7}{-10}$, and $-\dfrac{7}{10}$ are _____ in value.

3. The quotient of two numbers with different signs is _____.

4. The quotient of two numbers with same signs is _____.

EXAMPLES AND PRACTICE

Review this example for Objective A:

Divide signed numbers.

1. Find the quotient: $-20 \div (-5)$

Find the absolute values.

$|-20| = 20$ and $|-5| = 5$

Divide the absolute values $20 \div 5 = 4$

The numbers have the same sign, so the quotient is positive.

$-20 \div (-5) = 4$

---

2. Simplify: $\dfrac{-9}{18}$

$|-9| = 9$ and $|18| = 18$

$\dfrac{9}{18} = \dfrac{1}{2}$

The number have different signs, so the answer is negative.

$\dfrac{-9}{18} = -\dfrac{1}{2}$

Review this example for Objective B:

Solve applied problems involving the division of signed numbers.

3. The stock market changed -210 in three days. What is the average daily change?

$-\dfrac{210}{3} = -70$

Practice:

1. Find the quotient: $-36 \div (-9)$

2. Simplify: $\dfrac{-9}{27}$

3. The stock market changed -116 in four days. What is the average daily change?

ADDITIONAL EXERCISES
Objective A Divide signed numbers.

Find the quotient. Simplify.

1. $39 \div (-13)$

2. $-36 \div 3$

3. $-300 \div (-6)$

4. $246 \div (-1)$

5. $540 \div 10$

6. $-4.2 \div 3$

7. $-12 \div (-0.2)$

8. $\left(-\dfrac{3}{4}\right) \div \dfrac{6}{7}$

9. $\left(-\dfrac{8}{15}\right) \div \left(-\dfrac{9}{30}\right)$

10. $-70 \div 3\dfrac{1}{2}$

11. $\dfrac{-14}{-18}$

12. $\dfrac{8}{-20}$

13. $\dfrac{-13}{3}$

14. $\dfrac{-19}{-8}$

Objective B Solve applied problems involving the division of signed numbers.

Solve. Express each answer as a signed number.

15. Shawna scored 71 points in the last five games. How many points did she average over the five games?

16. The population of Cincinnati, OH decreased by 33,000 between 1990 and 2000. Find the average annual change in population.

17. Jimmy lost $3 on each of several rounds at the casino. His total loss was $57. How many rounds did he play?

18. A recipe calls for $1\dfrac{1}{4}$ cups of flour for one batch of cookies. How many batches can be made with 10 cups of flour?

Chapter 8 BASIC STATISTICS

8.1 Introduction to Basic Statistics

Learning Objectives
A   Find the mean, median, or mode(s) of a set of numbers.
B   Find the range of a set of numbers.
C   Solve applied problems involving basic statistics.

MATHEMATICALLY SPEAKING

In exercises 1-5 fill in the blank with the most appropriate term or phrase from the given list.

range	**weighted**	**mean**	**median**	**statistics**
arithmetic	**mode**	**algebra**		

1.   The branch of mathematics that deals with handling large quantities of information is called _____.

2.   The _____ is the difference between the largest number and the smallest number in a set of numbers.

3.   The _____(s) is the number (or numbers) occurring most frequently in a set of numbers.

4.   The sum of the numbers in a set divided by however many numbers are in the set is called the _____ mean.

5.   The middle number in a set of numbers arranged in numerical order is called the _____.

EXAMPLES AND PRACTICE

Review this example for Objective A:

Find the mean, median, or mode(s) of a set of numbers.

1. Find the mean, median, and mode of the set of numbers:

8, 2, 6, 4, 11, 5, 3, 8

Mean:

$$\frac{8+2+6+4+11+5+3+8}{8} = \frac{47}{8} = 5.875$$

Median:

Arrange the numbers in order from smallest to largest.

2 3 4 5 6 8 8 11

No single number is in the middle. In this case, the median is the mean of the two middle numbers.

2 3 4 $\boxed{5 \ 6}$ 8 8 11

$$\frac{5+6}{2} = 5.5$$

Mode: 8 occurs twice. It is the mode

Practice:

1. Find the mean, median, and mode of the set of numbers:

9, 3, 7, 5, 12, 6, 4, 9

Review this example for Objective B:

Find the range of a set of numbers.

2. Find the range of the numbers 4, 14, 3, 6, 10 and 3.
The largest number is 14.
The smallest number is 3.
The range is $14 - 4 = 11$

2. Find the range of the numbers 5, 15, 4, 7, 10 and 4.

Name:

Instructor:

Date:

Section:

Review this example for Objective C:

Solve applied problems involving basic statistics.

3. The SuperWash Car Wash had its grand opening. What is the mean, median, mode and range of the total cars washed in the last 10 days?

Mon	Tues	Wed	Thurs	Fri
30	35	27	53	60
Sat	Sun	Mon	Tues	Wed
50	35	33	37	30

Mean:

$$\frac{30 + 35 + 27 + 53 + 60 + 50 + 35 + 33 + 37 + 30}{10}$$

$$= \frac{390}{10} = 39$$

Median:

27 30 30 33 $\boxed{35\ 35}$ 37 50 53 60

The median is 35.

Mode: 30 and 35 are the modes.

Range: $60 - 27 = 33$

3. The salaries of the executives of a small company are listed below.

George	$78,000
Andrea	$82,000
Sarah	$39,000
Ahmed	$67,000
Tessa	$82,000

Determine the mean, median, mode and range of the salaries of the executives.

ADDITIONAL EXERCISES
Objective A Find the mean, median, or mode(s) of a set of numbers.

Find the mean, median, and mode(s) of each set of numbers. Round to the nearest tenth, where necessary.

1. 343, 392, 371, 371, 382

2. 35, 48, 21, 29

3. $36.25, $48.36, $48.36, $27.31

4. $6\frac{1}{2}, 4\frac{1}{4}, 3\frac{1}{3}, 9\frac{2}{3}, 4\frac{1}{4}$

5. $-10, -6, -2, -4, -4, -1$

Objective B Find the range of a set of numbers.

Find the range of each set of numbers.

6. 343, 392, 371, 371, 382

7. 35, 48, 21, 29

8. $36.25, $48.36, $48.36, $27.31

9. $6\frac{1}{2}, 4\frac{1}{4}, 3\frac{1}{3}, 9\frac{2}{3}, 4\frac{1}{4}$

10. $-10, -6, -2, -4, -4, -1$

Objective C Solve applied problems involving basic statistics.

Solve.

11. A college basketball team scored 85, 93, 80, 85, 92, 91, and 90 in its first seven games. What was the mean number of points scored?

12. A student had the following test scores: 72, 72, 45, 78, and 89. What was the student's average?

13. The high temperatures for seven days were: 79°, 83°, 88°, 73°, 75°, 75°, 80°. For these temperatures, what was the median and mode?

14. The median annual salary of 121 members of a department is $42,650. How many of the members have a salary that is higher than this salary?

15. A student's grade for a term is given as follows.

Grades	Credits	Grade Equivalent
A	2	4
B	4	3
C	5	2

Compute the students GPA for the term.

Chapter 8 BASIC STATISTICS

8.2 Tables and Graphs

Learning Objectives
A   Solve applied problems involving tables.
B   Solve applied problems involving graphs.

MATHEMATICALLY SPEAKING

In exercises 1-5 fill in the blank with the most appropriate term or phrase from the given list.

line graph	circle graph	heading	rows	columns
histogram	bar graph	graph	pictograph	table

1. A _____ is a picture or diagram of data.

2. In a table, _____ run horizontally.

3. On a _____, images of people, books, coins, and so on are used to represent quantities.

4. A _____ is a graph of a frequency table.

5. A _____ resembles a pie, representing the whole, that has been cut into slices, representing the parts.

EXAMPLES AND PRACTICE
Review this example for Objective A:

Solve applied problems involving tables.

1. The table below shows the number of cars that were parked on different streets at three different times during the day.

Parking Pattern			
Street	**9-10 am**	**12-1 pm**	**4-5 pm**
Oak	15	9	6
Maple	12	14	17
Hickory	18	20	22
Elm	6	4	5
Chestnut	4	3	6
Cherry	9	8	10
Ash	11	14	12

a. How many cars were parked on Hickory between 4 and 5?

b. What was the total number of cars parked on Chestnut street during the three time periods?

Solution

a. There were 22 cars on Hickory between 4 and 5.

b. There were 4 + 3 + 6 = 13 cars parked on Chestnut Street during the time periods.

Practice:

1. The table below shows the number of cars that were parked on different streets at three different times during the day.

Parking Pattern			
Street	**9-10 am**	**12-1 pm**	**4-5 pm**
Oak	15	9	6
Maple	12	14	17
Hickory	18	20	22
Elm	6	4	5
Chestnut	4	3	6
Cherry	9	8	10
Ash	11	14	12

a. How many cars were parked on Maple between 12 and 1?

b. What was the total number of cars parked on Oak street during the three periods?

Review this example for Objective B:

Solve applied problems involving graphs.

2. The following circle graph shows how Michelle spends her money each month.

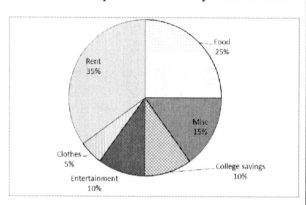

a. What percent of her money does Michelle use for rent?

b. If Michelle earns $2300 per month, how much of that does she save for college?

Solution

a. Michelle uses 35% of her money for rent.

b. 10% of 2,300 is used for college

$0.10(2,300) = \$230$

Michelle saves $230 per month for college.

2. The following circle graph shows how Daniel spends his money each month.

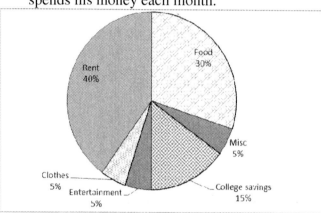

a. What percent of his money does Daniel spend on entertainment?

b. If Daniel earns $4300 per month, how much does he spend on food?

ADDITIONAL EXERCISES
Objective A Solve applied problems involving tables.

Solve.

1. The following table lists the number of visitors at several tourist attractions.

Attraction	Number of Visitors, in millions
Walt Disney World (FL)	14.0
Seaworld (FL)	5.2
Hershey Park (PA)	2.6
Knott's Berry Farm (CA)	3.5
Universal Studios (CA)	4.6

    a. How many people visited Knott's Berry Farm?

    b. By how much does the highest number of visitors at any tourist attraction exceed the lowest number of visitors?

2. The following table lists the number of years each area was a territory before it became a state.

Area	Years
Mississippi	19
Indiana	16
Michigan	31
Kansas	6
Florida	23

    a. How long was Indiana a territory before it became a state?

    b. Find the mean number of years it took these states to become a state.

3. The following table lists the number of electoral votes for several states.

State	Number of Votes
Washington	11
New Mexico	5
Kentucky	8
South Dakota	3
Georgia	15

    a. How many more electoral votes does Georgia have than Kentucky?

    b. What two states have the same number of electoral votes as Washington?

Name: Date:

Instructor: Section:

Objective B Solve applied problems involving graphs.

Solve.

4. This pictograph shows the number of consumer complaints to the Department of Transportation in a recent year against U. S. Airlines in several categories.

✈ Consumer Complaints to D.O.T. Against Airlines ✈	
Flight Problems	▰▰▰▰▰▰▰▰▰▰▰▰▱
Customer Service	▰▰▰▰▰▰▰▱
Baggage	▰▰▰▰▰▰▰▰▰▰▱
Refunds	▰▰▰▱
Fares	▰▱
▰ = 100 complaints	

a. Which category had the fewest number of complaints?

b. In which category were there about 750 complaints?

5. This pictograph shows the population for several countries in a recent year.

Population	
Luxembourg	♟♟♟♟♟
Qatar	♟♟♟♟♟♟♟♟♟
St. Lucia	♟♟
Malta	♟♟♟♟
Iceland	♟♟♟
Solomon Islands	♟♟♟♟♟♟
♟ = 100,000	

a. Which country has a population of about 400,000?

b. What is the population of Qatar?

6. The following bar graph shows the percent of adults who attended an artistic activity in the past 12 months (excludes elementary and high school performances).

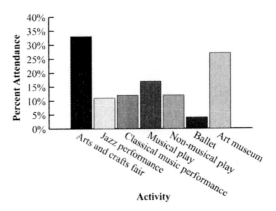

a. Approximately what percent of adults went to an arts and crafts fair in the past 12 months?

b. A company has 400 employees. How many would be expected to have gone to a musical play in the last year?

7. The following circle graph shows the distribution of expenditures on fishing related recreation in a recent year.

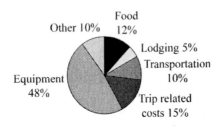

a. What percent of fishing related recreation expenditures was spent on lodging?

b. If $44 million was spent on fishing related recreation, how much of that was on food?

8. The following circle graph shows the types of investments by WealthAid Securities.

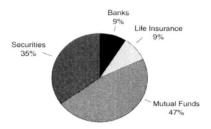

a. What percent of investments were not in banks and life insurance?

b. If you contributed $6,000 to this firm, how would it be allocated over all of the different investments?

9. The line graph below shows the personal savings as a percent of disposable personal income in the U.S. for various years.

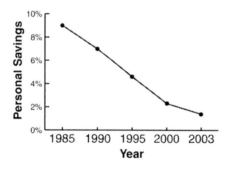

a. In what year was the percent of disposable personal income saved the greatest?

b. In what year was the percent of disposable personal income saved about 1.5%?

10. The line graph below shows the average temperature in Albany, NY for various months.

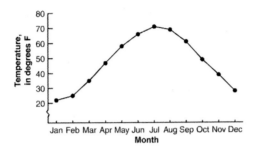

a. What was the average temperature in September?

b. In what month was the average temperature about 40° F?

c. Approximately how much warmer was the hottest month's temperature than the coldest?

Chapter 9 MORE ON ALGEBRA

9.1 Solving Equations

Learning Objectives
A   Solve equations involving signed numbers.
B   Solve equations with more than one step.
C   Solve applied problems involving equations with signed numbers or more than one step.

EXAMPLES AND PRACTICE

Review this example for Objective A: | **Practice:**

Solve equations involving signed numbers.

1. Solve and check: $y + 7 = 2$

Solution

$$y + 7 = 2$$
$$y + 7 - 7 = 2 - 7$$
$$y + 0 = -5$$
$$y = -5$$

1. Solve and check: $y + 8 = 3$

Check $y + 7 = 2$
$$-5 + 7 = 2$$
$$2 = 2 \text{ true}$$

The solution is –5.

---

2. Solve and check: $n - 7 = -15$

2. Solve and check: $n - 9 = -21$

Solution
$$n - 7 = -15$$
$$n - 7 + 7 = -15 + 7$$
$$n + 0 = -8$$
$$n = -8$$

Check $y - 7 = -15$
$$-8 - 7 = -15$$
$$-15 = -15 \text{ true}$$

The solution is –8.

3. Solve and check: $-8n = 32$

Solution $-8n = 32$

$$\frac{-8n}{-8} = \frac{32}{-8}$$

$$n = -4$$

Check $-8n = 32$

$$-8(-4) = 32$$

$$32 = 32 \quad \text{true}$$

The solution is –4.

3. Solve and check: $-8n = 48$

Review this example for Objective B:

Solve equations with more than one step.

4. Solve and check: $4y + 7 = -1$

Solution $4y + 7 = -1$

$$4y + 7 - 7 = -1 - 7$$

$$4y = -8$$

$$\frac{4y}{4} = \frac{-8}{4}$$

$$y = -2$$

Check $4y + 7 = -1$

$$4(-2) + 7 = -1$$

$$-8 + 7 = -1$$

$$-1 = -1$$

The solution is –2.

4. Solve and check: $4y + 7 = -17$

5. Solve and check: $\dfrac{a}{5} - 3 = 1$

Solution $\quad \dfrac{a}{5} - 3 = 1$

$$\dfrac{a}{5} - 3 + 3 = 1 + 3$$

$$\dfrac{a}{5} = 4$$

$$5 \cdot \dfrac{a}{5} = 5 \cdot 4$$

$$a = 20$$

Check $\dfrac{a}{5} - 3 = 1$

$$\dfrac{20}{5} - 3 = 1$$

$$4 - 3 = 1$$

$$1 = 1$$

The solution is 20.

5. Solve and check: $\dfrac{a}{4} - 7 = 1$

Review this example for Objective C:

Solve applied problems involving equations with signed numbers or more than one step.

6. A taxi charges $8 plus $1.10 per mile driven. How many miles were driven if the total cost of the ride was $69.60?

Solution $\quad 8 + 1.10x = 69.60$

$$8 + 1.10x = 69.60$$

$$8 - 8 + 1.10x = 69.60 - 8$$

$$1.10x = 61.60$$

$$\dfrac{1.10x}{1.10} = \dfrac{61.60}{1.10}$$

$$x = 56$$

The number of miles driven was 56.

6. An appliance repair technician charges $45 plus $35 per hour. How many hours did the technician work if the total bill was $167.50?

ADDITIONAL EXERCISES

Objective A Solve equations involving signed numbers.

Solve and check.

1. $x + 37 = 98$

2. $y - 53 = 141$

3. $59 + a = -123$

4. $-72 + t = -40$

5. $a + \dfrac{5}{6} = -\dfrac{1}{2}$

6. $-25a = -200$

7. $-15y = 96$

8. $\dfrac{x}{3} = -48$

Objective B Solve equations with more than one step.

Solve and check.

9. $5a - 4 = 26$

10. $8r + 16 = -48$

11. $-10x - 41 = 69$

12. $-2y + 7 = 7$

13. $11 = 4x + 6$

14. $3x - 8 = 12$

15. $6 - \dfrac{1}{2}t = 3$

16. $\dfrac{z}{5} - 13 = 0$

Objective C Solve applied problems involving equations with signed numbers or more than one step.

Write an equation. Solve and check.

17. If you make a deposit of $85, your account balance would be $415. What is the present balance of your account?

18. Cameron's annual salary is twice that of Joy's salary. If Cameron's salary is $65,380, what is Joy's salary?

19. On a segment of a flight, a plane descended from altitude of 22,000 feet to 18,750 feet. How far had the plane descended?

20. Jeremiah purchased a car for $20,000. It decreased in value an average of $1,600 per year. If he sold the car for $800, how long did he own it?

Chapter 9 MORE ON ALGEBRA

9.2 More on Solving Equations

EXAMPLES AND PRACTICE

Review this example for Objective A: | **Practice:**

Combine like terms.

1. Combine $8x + 7x$

Solution

$8x + 7x = (8 + 7)x$
$\qquad\quad = 15x$

1. Combine $6x + 4x$

Review this example for Objective B:

Solve equations involving like terms.

2. Solve: $4t - 6 = 5t + t - 18$
Solution
$\qquad 4t - 6 = 5t + t - 18$
$\qquad 4t - 6 = 6t - 18$
$\quad 4t - 4t - 6 = 6t - 4t - 18$
$\qquad\quad -6 = 2t - 18$
$\quad -6 + 18 = 2t - 18 + 18$
$\qquad\quad 12 = 2t$
$\qquad \dfrac{12}{2} = \dfrac{2t}{2}$
$\qquad\quad 6 = t$

2. $3t - 5 = 7t + t - 15$

Review this example for Objective C:

Solve equations involving parentheses.

3. Solve: $4w = 6(w + 6) + 8$

Solution

$$4w = 6(w + 6) + 8$$
$$4w = 6w + 36 + 8$$
$$4w = 6w + 44$$
$$4w - 44 = 6w + 44 - 44$$
$$4w - 44 = 6w$$
$$4w - 4w - 44 = 6w - 4w$$
$$-44 = 2w$$
$$\frac{-44}{2} = \frac{2w}{2}$$
$$-22 = w$$

3. Solve: $5t = 3(t + 4) + 6$

Review this example for Objective D:

Solve applied problems involving equations with like terms or parentheses.

4. Twice the sum of a number and 5 is 48. What is the number?

Solution
Write and equation and solve.
$2(n + 5) = 48$
$$2n + 10 = 48$$
$$2n + 10 - 10 = 48 - 10$$
$$2n = 38$$
$$\frac{2n}{2} = \frac{38}{2}$$
$$n = 19$$

4. Twelve less than four times a number is the same as six times the number. Find the number.

ADDITIONAL EXERCISES
Objective A Combine like terms.

Simplify.

1. $6x + 5x$ 2. $9y - 7y$

3. $8a + 5 + 4a$ 4. $7 - 2w + 3w$

Objective B Solve equations involving like terms.

Solve and check.

5. $3x + 5x = 48$ 6. $9t - 16t = -49$

7. $6x + 5 - 2x = -19$ 8. $4y + 2y - 7 = 3y + 11$

Objective C Solve equations involving parentheses.

Solve and check.

9. $2(n + 6) = -16$ 10. $7(x - 2) = 30$

11. $4(y - 3) = 3y$ 12. $6n + 2(n - 3) = 18$

Objective D Solve applied problems with like terms or parentheses.

Write an equation. Solve and check.

13. Four tickets to the Broadway show The Color Purple totaled $448. A service charge of $4 was added to each ticket. What was the price of each ticket before the service charge?

14. A 450-m fence is divided into three sections. The second section is twice as long as the first. The third section is three times as long as the second. Find the lengths of the sections.

15. An NBA basketball court is rectangular with a perimeter of 288 ft. The length is 44 ft longer than the width. Find the length and width of the court.

Chapter 9 MORE ON ALGEBRA

9.3 Using Formulas

Learning Objectives
A Translate a rule to a formula.
B Evaluate formulas.
C Solve applied problems involving formulas.

EXAMPLES AND PRACTICE

Review this example for Objective A:

Translate a rule to a formula.

1. To convert cricket chirps to degrees Fahrenheit, count the number of chirps in 14 seconds and add 40.

Solution

Temperature $= c + 40$

Practice:

1. To convert cricket chirps to degrees Celsius, count the number of chirps in 25 seconds, divide by three and then add 4.

Review this example for Objective B:

Evaluate formulas.

2. The formula for the volume of a cylinder is $V = \pi r^2 h$. What is the volume of a cylinder with a radius of 4 inches and a height of 6 inches? Round to the nearest whole number.
Solution
$V = \pi r^2 h$
$V \approx (3.14)(4)^2(6)$
$V \approx (3.14)(16)(6)$
$V \approx 301.44$
The volume is about 301 cubic inches.

2. The formula for the volume of a cylinder is $V = \pi r^2 h$. What is the volume of a cylinder with a radius of 6 inches and a height of 8 inches? Round to the nearest whole number.

Review this example for Objective C:

Solve applied problems involving formulas.

3. The perimeter of a rectangle is given by the formula $P = 2l + 2w$. If the perimeter is 52 cm and the length is 8 cm greater than the width, find the width and the length.

Solution

$$P = 2l + 2w$$

$$52 = 2(w + 8) + 2w$$

$$52 = 2w + 16 + 2w$$

$$52 = 4w + 16$$

$$52 - 16 = 4w + 16 - 16$$

$$36 = 4w$$

$$\frac{36}{4} = \frac{4w}{4}$$

$$9 = w$$

The width is 9 cm and the length is $(9 + 8) = 17$ cm.

3. The formula for the area of a trapezoid is $A = \frac{1}{2}h(b + B)$. Find the area if $h = 5$ m, $b = 3$ m, and $B = 7$ m.

ADDITIONAL EXERCISES
Objective A Translate a rule to a formula.

Convert to a formula.

1. To find the area A of triangle, take $\frac{1}{2}$ of the product of the base b and the height h.

2. The find the perimeter P of a rectangle, take the sum of twice the length l and twice the width w.

3. The circumference C of a circle is π times its diameter, d.

4. The area A of a square is the square of s, the length of one of its sides.

5. The area A of a semicircle is the product of $\frac{1}{2}\pi$ and the square of the radius r.

Objective B Evaluate formulas.

Evaluate each formula for the given quantity.

6. $A = lw$
Given $l = 5$ ft and $w = 3$ ft, find A.

7. $C = \dfrac{5F - 160}{9}$
Given $F = 77°$, find C.

8. $A = P + I$
Given $P = \$700$ and $I = \$31.50$, find A.

9. $C = 2\pi r$
Given $r = 4$ cm, find C.

10. $A = s^2 h$
Given $s = 3$ in. and $h = 7$ in., find A.

Objective C Solve applied problems involving formulas.

Solve.

11. The volume of a right circular cylinder is $V = \pi r^2 h$. Find the volume of a cylinder having $r = 2$ in. and $h = 5$ in.

12. If $2,000 is invested in an account paying 8% annual interest compounded quarterly, the amount in the account at the end of year t is given by $A = 2000(1.02)^{4t}$. How much is in the account at the end of the first year?

13. The formula $V = \dfrac{5600}{m}$ gives the volume of air (in ft^3) in a container when a piston of mass m attached to the container compresses the air inside. What is the volume of air in the container when the piston's mass is 8 kg?

Chapter 10 MEASUREMENT AND UNITS

10.1 U.S. Customary Units

Learning Objectives
A   Change a measurement from one U.S. customary unit to another.
B   Add or subtract measurements expressed in U.S. customary units.
C   Solve applied problems involving U.S. customary units.

MATHEMATICALLY SPEAKING

In exercises 1-5 fill in the blank with the most appropriate term or phrase from the given list.

**smaller   pound   gallon   weight   larger   numerator   denominator
unit factor   sign   length   unit**

1. In addition or subtraction problems, only quantities having the same _____
    can be added or subtracted.

2. When we change a large unit to a small unit, the numerical part of the answer is
    _____ than the original number.

3. In the U.S. customary system, the yard is a unit of _____.

4. In the U.S. customary system, the _____ is a unit of
    capacity.

5. When converting from one unit to another unit, multiply the original measure by the unit
    factor that has the desired unit in its _____.

EXAMPLES AND PRACTICE

Review this example for Objective A:

Change a measurement from one U.S. customary unit to another.

1. 7 qt = _____ pt

Solution

We want to change from quarts to pints, so we use the unit factor $\dfrac{2\text{ pt}}{1\text{ qt}}$

$7\text{qt} = 7\,\cancel{\text{qt}} \cdot \dfrac{2\text{ pt}}{1\,\cancel{\text{qt}}}$

$\phantom{7\text{qt}} = 14\text{ pt}$

Practice:

1. 36 yd = _____ ft

Review this example for Objective B:

Add or subtract measurements expressed in U.S. customary units.

2. Find the sum: 8 ft 7 in
                       + 3 ft 9 in

Solution

  8 ft 7 in   Start with inches. Add.
+ 3 ft 9 in
      16 in

  1 ft   16 inches = 1 ft 4 inches
  8 ft 7 in
+ 3 ft 9 in
    $\cancel{16}$ in 4 in

  1 ft
  8 ft 7 in
+ 3 ft 9 in
 12 ft 4 in

2. Find the sum: 7 ft 8 in
                       + 4 ft 5 in

Review this example for Objective C:

Solve applied problems involving U.S. customary units.

3. A scarf manufacturer uses 30 yards of yarn to make a scarf. How many feet of yarn are used to make the scarf?

    Solution

$$30 \text{ yd} = 30 \ \cancel{\text{yd}} \cdot \frac{3 \text{ ft}}{1 \ \cancel{\text{yd}}}$$

$$= 90 \text{ ft}$$

3. A fence on a ranch measures a total of 2 miles long. Determine the length of the fence in feet.

ADDITIONAL EXERCISES

Objective A Change a measurement from one U.S. customary unit to another.

Write each rate in simplest form.

1. 3 yd = _____ ft

2. 4 lb = _____ oz

3. 2.75 ton = _____ lb

4. 30 in. = _____ ft

5. 14 pt = _____ qt

6. 4 wk = _____ days

7. $3\frac{1}{2}$ gal = _____ pt

8. $\frac{1}{10}$ ton = _____ oz

9. 60 hr = _____ day

10. 500 sec = _____ min _____ sec

Objective B Add or subtract measurements expressed in U.S. customary units.

Compute.

11.   8 lb 6 oz
   − 5 lb 9 oz

12.   6 hr 15 min
   − 2 hr 40 min

13. 5 yr – 1 yr 8 mo   **14.** 3 ft 4 in. – 10 in.

15. 4 pt 10 fl oz + 2 pt 14 fl oz

Objective C Solve applied problems involving U.S. customary units.

Solve.

16. A puppy has gained 2.5 pounds in the last three weeks.
   a. How many ounces has the puppy gained?

   b. If the puppy weighed 36 ounces three weeks ago, how many pounds does it weigh now?

17. Martha's peach pie uses 2 pounds of peaches. How many ounces of peaches is this?

18. The height of the Statue of Liberty from the base to the torch is 151 feet 1 inch. What is the height in inches?

Chapter 10 MEASUREMENT AND UNITS

10.2 Metric Units and Metric/U.S. Customary Unit Conversions

Learning Objectives
A   Identify units in the metric system.
B   Change a measurement from one metric unit to another.
C   Change a measurement from a metric unit to a U.S. customary unit, and vice versa.
D   Solve applied problems involving metric units or U.S. customary units.

MATHEMATICALLY SPEAKING
In exercises 1-4 fill in the blank with the most appropriate term or phrase from the given list.

   kilo-   centi-   milli-   weight   deci-   quart   liter   length   capacity

1.   In the metric system, the liter is a unit of _____.

2.   The prefix _____ means one-thousandth.

3.   The prefix _____ means one-hundredth.

4.   The prefix _____ means one thousand.

Name: Date:
Instructor: Section:

EXAMPLES AND PRACTICE

Review this example for Objective A:	**Practice:**
Identify units in the metric system.	
1. An Olympic swimming pool is 50 meters long. Identify the metric unit.	1. The Peterson family traveled 3,200 km. Identify the metric unit.
Solution	
50 meters would be length	

Review this example for Objective B:

Change a measurement from one metric unit to another.

2. $800 \text{ m} = \_\_\_\_ \text{ km}$

Method 1

$$800 \text{ m} = 800 \text{ m} \cdot \frac{1 \text{ km}}{1000 \text{ m}}$$

$$= 0.8 \text{ km}$$

Method 2

Move the decimal point three 3 places to the left.

$$800 = \underset{\substack{\text{3 decimal} \\ \text{places}}}{.800} = 0.8 \text{ km}$$

2. $650 \text{ m} = \_\_\_\_ \text{ km}$

---

3. Express 4 kL in milliliters

Method 1

$$4 \text{ kL} = 4 \text{ kL} \cdot \frac{1000 \text{ L}}{1 \text{ kL}} \cdot \frac{1000 \text{ mL}}{1 \text{ L}}$$

$$= 4,000,000 \text{ mL}$$

Method 2

Move the decimal point 6 places to the right.

kL hL daL L dL cL mL

$$4 = 4,000,000 \text{ mL}$$

3. Express 7 hL in milliliters

Review this example for Objective C:

Change a measurement from a metric unit to a U.S. customary unit, and vice versa.

4. Express 7 qt in liters.

Solution

$$7 \text{ qt} \approx 7 \cancel{\text{qt}} \cdot \frac{1 \text{ L}}{1.1 \cancel{\text{qt}}}$$

$$\approx 6.4 \text{ L}$$

4. Express 7 pt in liters.

Review this example for Objective D:

Solve applied problems involving metric units or U.S. customary units.

5. A doctor prescribes 50 mg of Atenolol every 12 hours. How many grams of medication would a patient receive in one day?

Solution
50(2) = 100 mg total in a day

kg hg dag G dg cg mg

Move the decimal point 3 places left.
100 mg = 0.1 grams
The patient would receive 0.1 grams per day.

5. A science teacher is preparing an experiment. The experiment calls for 65 deciliters of chemicals, how many milliliters is this?

Name: Date:
Instructor: Section:

Objective A Identify units in the metric system.

State whether each quantity given is a measure of length, weight, or capacity.

1.   A toy rocket reached a maximum height of 12 meters.

2.   A gasoline tank can hold 80 liters gasoline.

3.   Our summer vacation covered 5,100 kilometers.

4.   Gary claims that he is 13.6 kilograms too heavy.

5.   The lab assistant had 1.3 grams of the substance.

Objective B Change a measurement from one metric unit to another.

Change each quantity to the indicated unit.

6.   8 m = _____ mm   7.   2700 ml = _____ L

8.   6.1 kg = _____ g   9.   1300 mg = _____ g

10. 4.875 mm = _____ cm   11. 5 L = _____ ml

Objective C Change a measurement from a metric unit to a U.S. customary unit, and vice versa.

Change each quantity to the indicated unit. If needed, round the answer to the nearest tenth of the unit.

12. 18 in. ≈ _____ cm   13. 400 km ≈ _____ mi

14. 8 ft ≈ _____ m   15. 700 g ≈ _____ lb

16. 5 gal ≈ _____ L

17. 10 kg ≈ _____ lb

18. 2 L ≈ _____ qt

19. 150 mi ≈ _____ km

Objective D Solve applied problems involving metric units or U.S. customary units.

Solve.

20. An aquarium holds 400 deciliters of water.

    a. How many liters of water does the aquarium hold?

    b. How many gallons does the aquarium hold?

21. Daria is driving and sees a road sign that says 450 kilometers. How far does she have to go in miles?

22. The height of the Statue of Liberty from the base to the torch is 151 feet 1 inch. What is the height in meters?

Chapter 11 BASIC GEOMETRY

11.1 Introduction to Basic Geometry

Learning Objectives
A   Identify basic geometric concepts or figures.
B   Find missing sides or angles.
C   Solve applied problems involving basic geometric concepts or figures.

MATHEMATICALLY SPEAKING

In exercises 1-10 fill in the blank with the most appropriate term or phrase from the given list.

radius	perpendicular	line segment	scalene	parallelogram
obtuse	acute	isosceles	trapezoid	supplementary
ray	parallel	vertical	diameter	complementary

1. A(n) _____ is a part of a line having two endpoints.

2. Two angles are _____ if the sum of their measures is 180°.

3. Lines that intersect to form right angles are called _____.

4. A(n) _____ triangle has two sides equal in length.

5. A(n) _____ triangle has no sides equal in length.

6. A(n) _____ is a line segment that passes through the center of a circle and has both endpoints on the circle.

7. A(n) _____ angle is an angle whose measure is more than 90° and less than 180°.

8. Angles with equal measure formed by two intersecting lines are called _____.

9. A(n) _____ is a quadrilateral with only one pair of opposite sides parallel.

10. A(n) _____ is a quadrilateral with both pairs of opposite sides parallel.

Name: Date:

Instructor: Section:

EXAMPLES AND PRACTICE

Review this example for Objective A: | **Practice:**

Identify basic geometric concepts or figures.

1. Find the measure of the angle complementary to 61°.

1. $\angle A$ and $\angle B$ are supplementary. If $m\angle A = 113°$, find the measure of $\angle B$.

Solution

$$m\angle A + m\angle B = 90°$$

$$m\angle A + 61° = 90°$$

$$m\angle A + 61° - 61° = 90° - 61°$$

$$m\angle A = 29°$$

Review this example for Objective B:

Find missing sides or angles.

2. Find the measure of the missing angle.

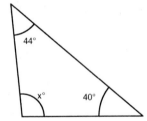

2. Find the measure of the missing angle.

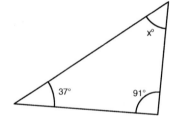

Solution

The sum of the measures of the angles is $180°$.

$$m\angle 1 + m\angle 2 + m\angle 3 = 180°$$

$$44° + 40° + x° = 180°$$

$$84° + x° = 180°$$

$$84° - 84° + x° = 180° - 84°$$

$$x° = 96°$$

Name: Date:
Instructor: Section:

Review this example for Objective C:	

Solve applied problems involving basic geometric concepts or figures.

3. A ceramic tile in the shape of a rectangle is being used in a new shower. 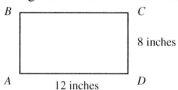 8 inches 12 inches What is the length of the side parallel to $\overline{CD}$? Solution The side parallel to $\overline{CD}$ is $\overline{AB}$. The length is 8 inches.	**3.** A ceramic tile in the shape of a rectangle is being used in a new shower. 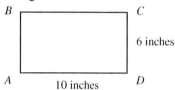 6 inches 10 inches What is the length of the side parallel to $\overline{BC}$?

Name: Date:
Instructor: Section:

ADDITIONAL EXERCISES
Objective A Identify basic geometric concepts or figures.

Identify each geometric object with the most appropriate name.

1.

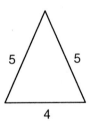

2.

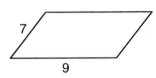

3.

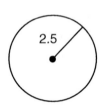

4.

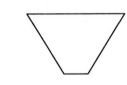

Solve.

5. Find the measure of the angle
that is supplementary to 48.5°.

6. ∠ABC is a straight angle. Find y.

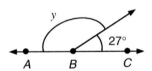

7. Find x.

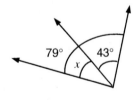

8. In the figure, $m\angle 1 = 35°$ and $m\angle 3 = 12°$.
Find $m\angle 2, m\angle 4,$
$m\angle 5,$ and $m\angle 6$.

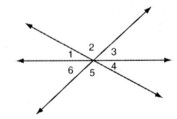

Name: Date:
Instructor: Section:

Objective B Find missing sides or angles.

9. Find the measure of the unknown
    angle.

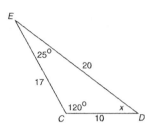

10. Find the measure of the unknown
    angle.

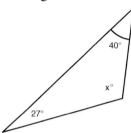

Objective C Solve applied problems involving basic geometric concepts or figures.

Solve. Use $\pi \approx 3.14$ *when needed.*

11. The circular top of a soda can has a diameter of 2.3 inches. What is the radius of
    the top of the can?

12. A triangular plot of land is such that two of the angles formed by the boundaries
    are complementary. If one of the two angles is 18°, what is the measure of the
    complementary angle?

13. In a quadrilateral, the measures of three of the four angles are 82°, 96°, and 123°.
    What is the measure of the remaining angle?

Chapter 11 BASIC GEOMETRY

11.2 Perimeter and Circumference

Learning Objectives
A Find the perimeter of a polygon or the circumference of a circle.
B Find the perimeter or circumference of a composite geometric figure.
C Solve applied problems involving perimeter or circumference.

MATHEMATICALLY SPEAKING
In exercises 1-4 fill in the blank with the most appropriate term or phrase from the given list.

**simple   rectangle   composite   circle   circumference   square   perimeter
length**

1. A formula for the circumference of a _____ is $C = 2\pi r$.

2. The distance around a circle is called its _____.

3. The _____ of a polygon is the distance around it.

4. Two or more basic geometric figures are combined in a _____ figure.

Name: Date:

Instructor: Section:

EXAMPLES AND PRACTICE

Review this example for Objective A: | **Practice:**

Find the perimeter of a polygon or the circumference of a circle.

1. Find the perimeter.

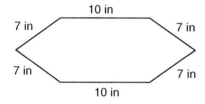

Solution

The distance around is the perimeter.

$P = 10$ in $+ 7$ in $+ 7$ in $+ 10$ in $+ 7$ in $+ 7$ in

$P = 48$ in

The perimeter of the shape is 48 inches.

2. Find the circumference. Use 3.14 for π.

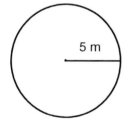

Solution

$C = 2\pi r$

$C \approx 2(3.14)(5)$

$C \approx 31.4$

The circumference is approximately 31.4 meters.

1. Find the perimeter.

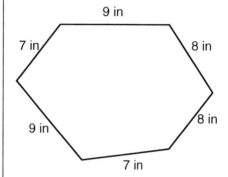

2. Find the circumference. Use 3.14 for π.

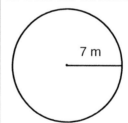

Name: Date:
Instructor: Section:

Review this example for Objective B:

Find the perimeter or circumference of a composite geometric figure.

3. Find the perimeter.

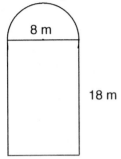

3. Find the perimeter.

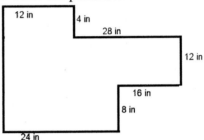

Solution
Circumference of circle:
The diameter is 8 m so the radius is 4 m.
$C = 2\pi r$

$C = 2(3.14)(4)$

$C = 25.12$
Since we only need half of a circle divide the
total circumference by 2. $\dfrac{25.12}{2} = 12.56$

We need the perimeter of only three sides of
the rectangle.
Perimeter of rectangle:
$P = 18\text{ m} + 8\text{ m} + 18\text{ m}$
$P = 44\text{ m}$

Total perimeter = 44 m + 12.56 m
Total perimeter = 56.56 m

Name: Date:
Instructor: Section:

Review this example for Objective C:

Solve applied problems involving perimeter or circumference.

4. New weather stripping needs to be installed around 120 portholes on a cruise ship. Each porthole has a diameter of 18 inches.
   a. How many inches of weather stripping is needed for each porthole?
   b. If weather stripping costs $0.15 per inch, what is the cost of weather stripping for all the windows?

Solution
a. Circumference of each porthole.

$C = 2\pi r$

$C = 2(3.14)(9)$

$C = 56.52$

56.52 inches of weather stripping are needed for each porthole.

b.   Total = porthole × # of holes × cost
     Total = 56.52(120)($0.15)
     Total = $1017.36

It will cost $1017.36 to replace the weather stripping.

4. The grazing pasture on Kevin's ranch is in need of new fencing. The pasture is a rectangle 120 yards long and 72 yards wide.
   a. What is the perimeter of the pasture?
   b. If fencing costs $35 per yard, what is the total cost of the fence?

Name: _____ Date: _____

Instructor: _____ Section: _____

ADDITIONAL EXERCISES

Objective A Find the perimeter of a polygon or the circumference of a circle.

Find the perimeter of each figure.

1.

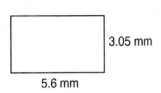

3.05 mm

5.6 mm

2.

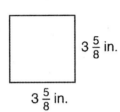

$3\frac{5}{8}$ in.

$3\frac{5}{8}$ in.

3.

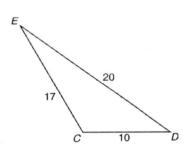

4.

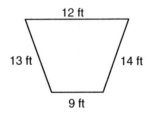

Find the circumference. Use $\pi \approx 3.14$.

5.

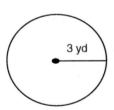

3 yd

6.

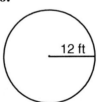

12 ft

7.

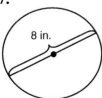

8 in.

8.

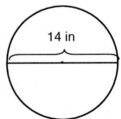

14 in

Objective B Find the perimeter or circumference of a composite geometric figure.

Find the perimeter of each figure. Use π ≈ 3.14 when needed.

9.

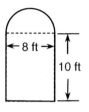

10.

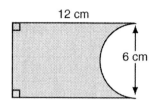

11.

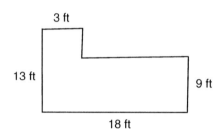

12.

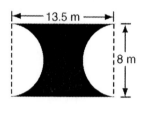

Objective C Solve applied problems involving perimeter or circumference.

Solve. Use π ≈ 3.14 when needed.

13. An NBA basketball court is rectangular with length 94 ft and width 50 ft. What is the perimeter of the court?

14. A wheel of bicycle is 24 inches in diameter. What is the circumference of the wheel?

15. The Pentagon in Arlington, Virginia, is a five-sided building with each side measuring 921 ft. What is the perimeter of the building?

Chapter 11 BASIC GEOMETRY

11.3 Area

Learning Objectives
A   Find the area of a polygon or a circle.
B   Find the area of a composite geometric figure.
C   Solve applied problems involving area.

MATHEMATICALLY SPEAKING
In exercises 1-4 fill in the blank with the most appropriate term or phrase from the given list.

trapezoid   volume   square inches   circle   meters   triangle   square   area

1.   The area of a(n) _____ is equal to one-half the product of the base and
     the height.

2.   The formula $A = \frac{1}{2}h(b+B)$ is used to find the area of a(n) _____.

3.   The area of a polygon is measured in square units, such as _____.

4.   The formula for the area of a(n) _____ is $A = \pi r^2$.

Name: Date:

Instructor: Section:

EXAMPLES AND PRACTICE

Review this example for Objective A: | **Practice:**

Find the area of a polygon or a circle.

1. Find the area.

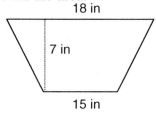

$$A = \frac{1}{2}h(b + B)$$

$$= \frac{1}{2} \cdot 7(18 + 15)$$

$$= \frac{1}{2} \cdot 7(33)$$

$$= \frac{231}{2} \text{ or } 115\frac{1}{2}$$

The area of the trapezoid is $115\frac{1}{2}$ square inches.

2. Find the area. Use 3.14 for π.

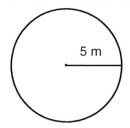

Solution

$A = \pi r^2$

$A \approx (3.14)(5)^2$

$A \approx (3.14)(25)$

$A \approx 78.5$

The area of the circle is approximately 78.5 square meters.

1. Find the area.

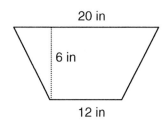

2. Find the area. Use 3.14 for π.

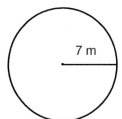

Name: Date:
Instructor: Section:

Review this example for Objective B:

Find the area of a composite geometric figure.

3. Find the area.

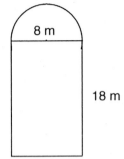

3. Find the area.

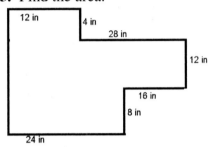

Solution
The radius of the circle is 4 m.
Area of an entire circle:

$A = \pi r^2$

$A \approx (3.14)(4)^2$

$A \approx (3.14)(16)$

$A \approx 50.24$

Area of half a circle: $\dfrac{50.24}{2} = 25.12$

Area of the rectangle:

$A = l \cdot w$

$A = 18 \cdot 8$

$A = 144$

Total area = area of half circle + area of rectangle

Total Area = 25.12 + 144 = 169.12 m^2

Review this example for Objective C:

Solve applied problems involving area.

4. Dennis is building a new outdoor deck. He is trying to determine the area of the deck in order to purchase materials. The plans call for a square deck with each side measuring 12.5 feet. What is the area of the deck?

Solution

$A = s^2$

$A = 12.5^2$

$A = 156.25$

The area of the deck is 156.25 square feet.

4. A local artist club is hosting a weekend day camp for kids. Each child will receive a square piece of art paper measuring 18 inches in length. What is the area of the paper?

ADDITIONAL EXERCISES

Objective A Find the area of a polygon or a circle.

Find the area of each figure. Use $\pi \approx 3.14$ *when needed.*

1.

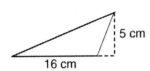

5 cm

16 cm

2.

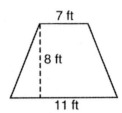

7 ft

8 ft

11 ft

3.

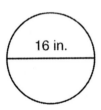

16 in.

4.

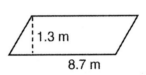

1.3 m

8.7 m

5.

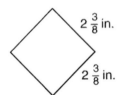

$2\frac{3}{8}$ in.

$2\frac{3}{8}$ in.

Objective B Find the area of a composite geometric figure.

Find the area of each figure or shaded region. Use $\pi \approx 3.14$ *when needed.*

6.

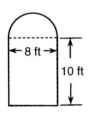

8 ft

10 ft

7.

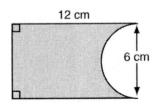

12 cm

6 cm

Name: Date:

Instructor: Section:

8.

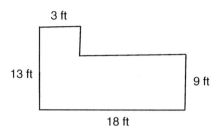

9.

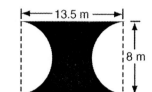

10.

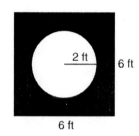

Objective C Solve applied problems involving area.

Solve. Use $\pi \approx 3.14$ *when needed.*

11. The diameter of a quarter is 2.5 cm. What is its area?

12. A rectangular garden is 8 ft long and $5\frac{1}{2}$ ft wide. What is the area of the garden?

13. A standard sheet of copy paper is 11 inches long and $8\frac{1}{2}$ inches wide. What is the area of the page?

14. Papa Henry's specialty pizza comes in two sizes. There is the 14-in. (one side) square pizza or a 14-in (diameter) round pizza for the same price. Which is the better buy?

15. The playing field inside a 400 m track is a rectangle with two semicircular ends. If the length of the rectangular portion is 84.4 m and the radius of the semicircular ends is 36.5 m, what is the area of the playing field?

Chapter 11 BASIC GEOMETRY

11.4 Volume

Learning Objectives
A   Find the volume of a geometric solid.
B   Find the volume of a composite geometric solid.
C   Solve applied problems involving volume.

MATHEMATICALLY SPEAKING
In exercises 1-4 fill in the blank with the most appropriate term or phrase from the given list.

> **circle   composite   volume   cube   rectangular solid   simple   cylinder**
> **area   sphere**

1.   The number of cubic units required to fill a three-dimensional figure is called its

     _____.

2.   A(n) _____ is a solid in which the bases are circles and are
     perpendicular to the height.

3.   Combining two or more basic solid figures results in a(n) _____
     geometric solid.

4.   The formula $V = lwh$ is used to find the volume of a(n) _____.

Name: _____   Date: _____

Instructor: _____   Section: _____

EXAMPLES AND PRACTICE

Review this example for Objective A:

Find the volume of a geometric solid.

1. Find the volume.

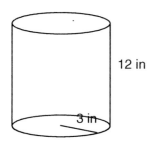

12 in

3 in

Solution

The radius of the cylinder is 3 inches.

The height is 12 inches.

$V = \pi r^2 h$

$\approx (3.14)(3)^2(12)$

$\approx (3.14)(9)(12)$

≈ 339.12

The area of the cylinder is approximately 339 cubic inches.

Review this example for Objective B:

Find the volume of a composite geometric solid.

2. Find the volume. The radius of the figure is 10 feet and the height of the cylinder is 15 feet.

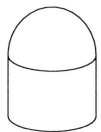

Practice:

1. Find the volume.

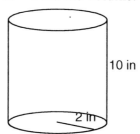

10 in

2 in

2. Find the volume. The radius of the figure is 8 feet and the height of the cylinder is 14 feet.

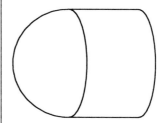

Solution
Find the volume of the cylinder and half a sphere.
Volume of the cylinder

$V = \pi r^2 h$

$\approx (3.14)(10)^2(15)$

$\approx (3.14)(100)(15)$

$\approx 4,710$

Volume of the sphere

$V = \dfrac{4}{3}\pi r^3$

$\approx \dfrac{4}{3}(3.14)(10)^3$

$\approx \dfrac{4}{3}(3,140)$

$\approx 4,186.67$

We only need the volume of half the sphere:

$\dfrac{4,186.67}{2} = 2,093.3$

Total Volume $= 4,710 + 2,093.3 = 6,803.3$

The volume of the figure is about 6,803.3 cubic feet.

Review this example for Objective C:

Solve applied problems involving volume.

3. The cylindrical tank of a gasoline truck is 30 feet long and has a diameter of 8 feet. Find the volume of the tanker truck.
Solution
The radius of the truck is 4 feet.
The height of the truck is 30 feet.

$V = \pi r^2 h$

$\approx (3.14)(4)^2(30)$

$\approx (3.14)(16)(30)$

≈ 1507.2

The volume of the tanker truck is approximately 1,507 cubic feet.

3. How many cubic feet of grain can be stored in a cylindrical grain silo that is 40 feet high and has a radius of 25 feet?

Name: Date:
Instructor: Section:

ADDITIONAL EXERCISES

Objective A Find the volume of a geometric solid.

Find the volume of each solid. Use $\pi \approx 3.14$ when needed.

1.

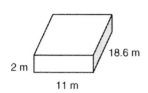

2.

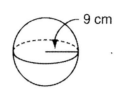

3.

4.

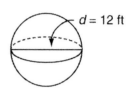

5.

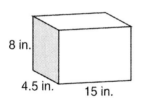

Objective B Find the volume of a composite geometric solid.

Find the volume of each solid. Use $\pi \approx 3.14$ when needed.

6.

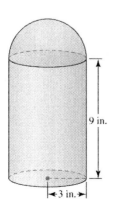

7.

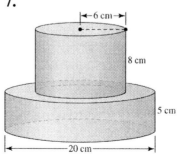

8.

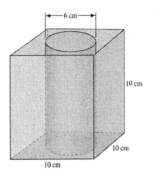

9.

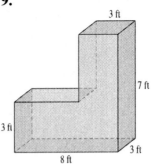

10.

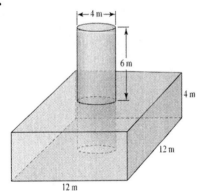

Objective C Solve applied problems involving volume.

Solve. Use $\pi \approx 3.14$ *when needed.*

11. A edge of a sugar cube measures $\frac{1}{4}$ in. What is the volume of the cube?

12. The radius of a standard-sized bowling ball is 4.29 inches. Find the volume of a standard-sized bowling ball. Round to the nearest hundredth cubic inch.

13. A concrete stave silo used for storing corn silage is a cylindrical structure with a hemispherical top. If the silo has a diameter of 16 ft and height 40 ft, how many cubic feet of silage can the silo store? Round to the nearest hundredth cubic ft.

14. The storage area of U-Move's compact moving truck resembles a rectangular solid having length 3.6 m, width 2.1 m, and height 1.9 m. What is the storage capacity of the van? Round to the nearest hundredth cubic meter.

15. The storage area of U-Move's Compact Moving Truck resembles a rectangular solid having length 12 ft, width 7 ft, and height $6\frac{1}{4}$ ft. A refrigerator having width 3 ft, length $2\frac{1}{2}$ ft, and height $5\frac{3}{4}$ ft is placed in the truck. What is the volume of the space remaining in the truck?

Chapter 11 BASIC GEOMETRY

11.5 Similar Triangles

Learning Objectives
A   Find the missing sides of similar triangles.
B   Solve applied problems involving similar triangles.

MATHEMATICALLY SPEAKING

In exercises 1-4 fill in the blank with the most appropriate term or phrase from the given list.

equal   shape   similar   area   in proportion   corresponding

1. In similar triangles, _____ sides are opposite angles with equal measure.

2. The symbol ~ is used to indicate that triangles are _____.

3. Corresponding sides of similar triangles are _____.

4. Similar triangles have the same _____ but not necessarily the same size.

Name: Date:
Instructor: Section:

EXAMPLES AND PRACTICE
  Review this example for Objective A: | **Practice:**

  Find the missing sides of similar triangles.

1. In the following diagram,
$\triangle DEF \sim \triangle ABC$. Find x.

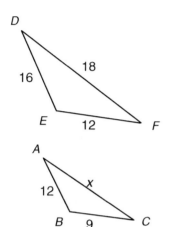

Solution

Write the ratios of the corresponding lengths.

$$\frac{DE}{AB} = \frac{DF}{AC} = \frac{EF}{BC}$$

$$\frac{16}{12} = \frac{18}{x} = \frac{12}{9}$$

Use any two proportions to solve.

$$\frac{16}{12} = \frac{18}{x}$$

$$16x = 12(18)$$

$$16x = 216$$

$$\frac{16x}{16} = \frac{216}{16}$$

$$x = 13.5$$

1. In the following diagram,
$\triangle ABC. \sim \triangle DEF$ Find y.

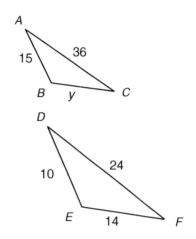

Review this example for Objective B:

Solve applied problems involving similar triangles.

2. Kyle is 5.8 feet tall. Late one afternoon while visiting a Lighthouse he noticed that his shadow was 9 feet long. At the same time a Lighthouse cast a shadow 108 feet long. What is the height of the lighthouse?

Solution

$$\frac{\text{height of Kyle}}{\text{shadow}} = \frac{\text{height of lighthouse}}{\text{lighthouse shadow}}$$

$$\frac{5.8}{9} = \frac{x}{108}$$

$$9x = 5.8(108)$$

$$9x = 626.4$$

$$\frac{9x}{9} = \frac{626.4}{9}$$

$$x = 69.6$$

The lighthouse is about 69.6 feet tall.

2. A palm tree with a height of 28 feet cast a shadow of 8 feet. How high is a telephone pole next to the tree, if the pole cast a shadow of 6 feet?

Name: Date:
Instructor: Section:

ADDITIONAL EXERCISES
Objective A Find the missing sides of similar triangles.

Each pair of triangles is similar. Find the length of each missing side.

1.

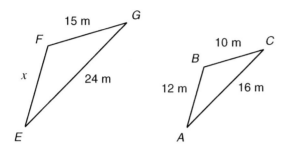

2.

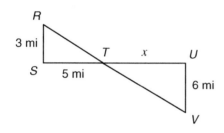

3.

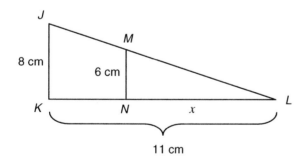

Name: Date:

Instructor: Section:

Objective B Solve applied problems involving similar triangles.

Solve. Assume the triangles are similar.

4. A 4 ft high fence casts a 3 ft shadow. How tall is a tree that cast a 27 ft shadow?

5. A flagpole cast a 58 ft shadow at the same time that a 6 ft man casts a 5 ft shadow. How tall is the flagpole?

6. An escalator that is 150 cm at its base takes people 132 cm high to the second floor of an office building. How high above the first floor will people be when the base has only covered 120 cm?

7. One way to measure the height of a building is to position a mirror on the ground so that the top of the building's reflection can be seen. Find the height *h*.

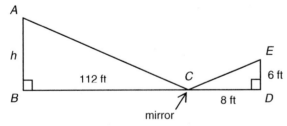

8. A power line pole 18 feet high casts a shadow 27 feet long. At the same time, the shadow of a building measures 78 feet long. How tall is the building?

Chapter 11 BASIC GEOMETRY

11.6 Square Roots and the Pythagorean Theorem

Learning Objectives
A Find the square root of a number.
B Find the unknown side of a right triangle, using the Pythagorean theorem.
C Solve applied problems involving a square root or the Pythagorean theorem.

MATHEMATICALLY SPEAKING

In exercises 1-4 fill in the blank with the most appropriate term or phrase from the given list.

squaring	**leg**	**hypotenuse**	**multiple**
Area of Three Squares	**prime**	**consecutive**	
perfect square	**doubling**	**Pythagorean theorem**	
square root			

1.   In a right triangle, the longest side is called the _____.

2.   The square of a whole number is said to be a(n) _____.

3.   The number 6 is the _____ of 36.

4.   If *a* and *b* are legs of a right triangle and *c* is the hypotenuse, then the
     _____ states that $a^2 + b^2 = c^2$.

EXAMPLES AND PRACTICE

Review this example for Objective A: | **Practice:**

Find the square root of a number.

1. Find the square root. $\sqrt{1600}$

Solution

$\sqrt{1600} = 40$ because $40 \cdot 40 = 1600$.

1. Find the square root. $\sqrt{2500}$

---

2. $\sqrt{52}$ lies between which two consecutive whole numbers?

Solution

Find two consecutive perfect squares that 52 lies between.

52 is more than 49 and less than 64, which are consecutive perfect squares.

$\sqrt{52}$ lies between 7 and 8.

2. $\sqrt{71}$ lies between which two consecutive whole numbers?

Review this example for Objective B:

Find the unknown side of a right triangle, using the Pythagorean theorem.

3. Find the missing length.

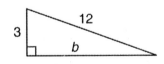

$a^2 + b^2 = c^2$

$3^2 + b^2 = 12^2$

$9 + b^2 = 144$

$9 - 9 + b^2 = 144 - 9$

$b^2 = 135$

$b = \sqrt{135}$

3. Find the missing length.

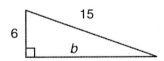

Name: Date:

Instructor: Section:

Review this example for Objective C:

Solve applied problems involving a square root or the Pythagorean theorem.

4. A 17-ft ladder leans against a building. The bottom of the ladder is 3 ft from the building. How high is the top of the ladder?

?   17 ft

3 ft

$$a^2 + b^2 = c^2$$

$$3^2 + b^2 = 17^2$$

$$9 + b^2 = 289$$

$$9 - 9 + b^2 = 289 - 9$$

$$b^2 = 280$$

$$b = \sqrt{280} \approx 16.7$$

The top of the ladder is about 16.7 feet high.

4. A cord reaches from the top of a 15-ft pole to a point on the ground 9 ft from the pole. How long is the cord?

183

ADDITIONAL EXERCISES
Objective A Find the square root of a number.

Find the square root.

1. $\sqrt{1}$

2. $\sqrt{25}$

3. $\sqrt{49}$

4. $\sqrt{144}$

Determine between which two consecutive whole numbers each square root lies.

5. $\sqrt{75}$

6. $\sqrt{7}$

7. $\sqrt{112}$

8. $\sqrt{40}$

Find each square root. Round to the nearest tenth if necessary.

9. $\sqrt{57}$

10. $\sqrt{21}$

11. $\sqrt{92}$

12. $\sqrt{175}$

Objective B Find the unknown side of a right triangle, using the Pythagorean theorem.

Find the missing length. Round to the nearest tenth, if needed.

13.

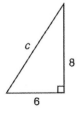

14.

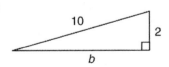

15.

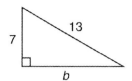

16.

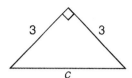

Objective C Solve applied problems involving a square root or the Pythagorean theorem.

Find the missing length. Round to the nearest tenth, if needed.

17. A baseball diamond is actually a square made up of the home plate and first, second, and third bases. The four bases are at the corners of the square, and the distance between any two consecutive bases is 90 feet. How far is it from first base to third base?

18. Two planes are heading toward the same airport. One plane is 8 miles east of the airport and the other is 12 miles north of the airport. If the planes are traveling at the same altitude, how far apart are the planes? Give answer to the nearest tenth of a mile.

19. Students on campus tend to congregate on the "yard", a rectangular region with grass of length 80 ft and width 64 ft. The college wants to create a diagonal walkway through the yard. To the nearest tenth, what would be the length of the walkway?

20. A television screen is measured by the length of the diagonal of the screen. A flat-screen TV measures 28 inches high and 48 inches wide. How is the TV advertised? Round to the nearest whole inch.

Chapter 1
WHOLE NUMBERS

1.1 Introduction to Whole Numbers
Mathematically Speaking
1. whole numbers
2. digits
3. odd
4. expanded form
5. rounded
6. standard form
7. place value

PRACTICE
1. a. ones
   b. thousands
2. fifty-six thousand, two hundred ten
3. 20,060,000
4. 20,000 + 1,000 + 400 + 50 + 9
5. 58,000
6. $200,000

ADDITIONAL EXERCISES
Objective A
1. 89,003
2. 326,590
3. Twenty-three thousand, eight hundred ninety
4. Twelve million, fifty-three thousand, seven
5. Ten thousand
6. Thousand
7. 46, 209
8. 12,700,000,192

Objective B
9. 900 + 80 + 6
10. 3000 + 0 + 10 + 7

Objective C
11. 12,800
12. 300,000

Objective D
13. 160 feet
14. 705,000 square miles

1.2 Adding and Subtracting Whole Numbers
Mathematically Speaking
1. Identity Property of Addition
2. addends
3. Commutative Property of Addition
4. subtrahend
5. Associative Property of Addition
6. difference

PRACTICE
1. 56,839
2. 391
3. 25,647; 20,900
4. 3,827; 4,000
5. $5,606
6. $858

Objective A
1. 5,034
2. 106,651
3. 56,839
4. 10,467
5. 11,514
6. 106,479
7. 1,676
8. $5,510
9. 175
10. 369,010
11. 5,186
12. 10,442
13. 18,000
14. 100,000

Objective B
15. 1,613,000 people
16. 13 mph
17. $450

1.3 Multiplying Whole Numbers
Mathematically Speaking
1. Distributive Property
2. addition
3. Identity Property of Multiplication
4. product

PRACTICE
1. 4,224
2. 149,736

3. 77,088
4. 960

Objective A
1. 1,592
2. 41,905
3. 31,220
4. 67,456
5. 104,962
6. 75,168
7. 802,386
8. 38,731,063
9. 12,000
10. 28,000
11. 100,000
12. 24,000

Objective B
13. 504 miles
14. 1,965 miles
15. $10,320
16. 1,575 words

1.4 Dividing Whole Numbers
Mathematically Speaking
1. quotient
2. divided
3. divisor
4. multiplication

PRACTICE
1. 508
2. 81 R 4
3. 52
4. 57 DVD's

Objective A
1. 50
2. 241
3. 121
4. 226
5. 15
6. 80
7. 117
8. 3027 R 13

9-12 are sample answers
9. 5,000
10. 300
11. 60
12. 400

Objective B
13. $109 per night
14. 9 minutes
15. 375 phone calls
16. 20 minutes

1.5 Exponents, Order of Operations, and Averages
Mathematically Speaking
1. product
2. adding
3. grouping
4. base

PRACTICE
1. $4^3 \cdot 5^2$
2. $5 \cdot 5 \cdot 5 \cdot 2 \cdot 2 \cdot 2 \cdot 2 \cdot 2 = 4,000$
3. 14
4. 143
5. 12
6. 90 ppg

Objective A
1. $4^3 \cdot 9^4$
2. $2^4 \cdot 11^5$
3. $2^3 \cdot 7^4 \cdot 13^2$
4. $5 \cdot 11^2 \cdot 17^3$
5. $5 \cdot 5 \cdot 3 \cdot 3 \cdot 3 = 675$
6. $8 \cdot 8 \cdot 10 \cdot 10 \cdot 10 = 64,000$

Objective B
7. 50
8. 38
9. 63
10. 15

Objective C
11. 27
12. 75

Objective D
13. 152
14. 84

1.6 More on Solving Word Problems
Mathematically Speaking
1. read
2. check
3. problem-solving strategy
4. clue words

PRACTICE
1. 307 tons
2. $1,275
3. $6,136
Objective A
1. $1827
2. 80 min
3. 308 ft/week
4. 3,000,000 contractions in a 30-day month
5. 80,000 books
6. 81 ft^2
7. $58
8. 285,931 students/year
9. 7 buses
10. 6 lbs/month

Chapter 2   FRACTIONS

2.1 Factors and Prime Numbers
Mathematically Speaking
1. common multiple
2. prime factorization
3. composite
4. least common multiple
5. factors
PRACTICE
1. 1, 2, 5, 10
2. a. composite
   b. prime
   c. composite
3. $42 = 2 \cdot 3 \cdot 7$
4. 90
5. every 24th customer
Objective A
1. 1, 2, 3, 4, 6, 8, 12, 16, 24, 48
2. 1, 3, 7, 9, 21, 63
3. 1, 2, 3, 4, 6, 9, 12, 18, 36
4. 1, 2, 3, 6, 17, 34, 51, 102
Objective B
5. composite; 2
6. prime
7. composite; 3
8. composite; 3
Objective C
9. $36 = 2^2 \cdot 3^2$

10. $216 = 2^3 \cdot 3^3$
11. $252 = 2^2 \cdot 3^2 \cdot 7$
12. $525 = 3 \cdot 5^2 \cdot 7$
Objective D
13. 28
14. 45
15. 84
16. 42
17. 24
18. 63
19. 40
20. 156
Objective E
21. yes
22. 12 laps
23. 20 days
24. 420 years
25. 4620 years

2.2 Introduction to Fractions
Mathematically Speaking
1. multiply
2. divide
PRACTICE
1. $\dfrac{5}{11}$
2. $\dfrac{47}{9}$
3. $3\dfrac{3}{4}$
4. sample answers: $\dfrac{12}{14}; \dfrac{18}{21}$
5. $\dfrac{30}{55} = \dfrac{6}{11}$
6. $\dfrac{3}{4} \boxed{>} \dfrac{11}{15}$
7. $\dfrac{10}{50} = \dfrac{1}{5}$
Objective A
1. $\dfrac{2}{3}$
2. $\dfrac{5}{6}$
3. improper fraction

4. improper fraction
5. mixed number

Objective B

6. $\dfrac{25}{1}$

7. $\dfrac{32}{9}$

8. $\dfrac{64}{5}$

9. $2\dfrac{3}{7}$

10. $6\dfrac{4}{9}$

11. 8

Objective C

12. $\dfrac{30}{54}$

13. $\dfrac{26}{39}$

14. $\dfrac{2}{3}$

15. $\dfrac{5}{11}$

16. $\dfrac{3}{10}$

17. $\dfrac{13}{4}$ or $3\dfrac{1}{4}$

Objective D

18. <
19. >

Objective E

20. $\dfrac{7}{22}$

21. $\dfrac{3}{4}$ mi to work

22. $\dfrac{4}{23}$

2.3 Adding and Subtracting Fractions
Mathematically Speaking

1. borrow
2. numerators
3. equivalent

PRACTICE

1. $\dfrac{14}{20} = \dfrac{7}{10}$

2. $\dfrac{11}{6} = 1\dfrac{5}{6}$

3. $1\dfrac{7}{12}$

4. $10\dfrac{17}{24}$

5. $24\dfrac{1}{2}$

6. $29 + 17 - 12 = 34; 33\dfrac{73}{132}$

7. $1\dfrac{11}{40}$ pounds

Objective A

1. $\dfrac{1}{4}$

2. $\dfrac{61}{60}$ or $1\dfrac{1}{60}$

3. $\dfrac{4}{45}$

4. $\dfrac{19}{60}$

5. $4\dfrac{5}{7}$

6. $16\dfrac{5}{6}$

7. $5\dfrac{7}{12}$

8. $3\dfrac{5}{6}$

9. 54
10. 42

Objective B

11. $\dfrac{7}{20}$ acre

12. $\dfrac{5}{12}$ cup

13. $71\dfrac{1}{3}$ in.

14. $18\dfrac{3}{4}$ in.

15. $\dfrac{11}{36}$

2.4 Multiplying and Dividing Fractions
Mathematically Speaking
1. improper fraction
2. divide
3. multiply
4. reciprocal

PRACTICE

1. $\dfrac{14}{27}$

2. $\dfrac{9}{25}$

3. $22\dfrac{2}{3};20$

4. $1\dfrac{1}{9}$

5. $\dfrac{37}{76}$

6. 24 volumes

Objective A

1. $\dfrac{3}{20}$

2. 38

3. $\dfrac{2}{3}$

4. $\dfrac{7}{24}$

5. $25\dfrac{1}{2}$

6. $\dfrac{29}{12}$ or $2\dfrac{5}{12}$

7. $4\dfrac{5}{6}$

8. $\dfrac{12}{35}$

9. $\dfrac{7}{9}$

10. $\dfrac{13}{70}$

11. 16
12. 9
Objective B
13. $45
14. $2750

15. $\dfrac{3}{10}$ lb

16. $\dfrac{7}{24}$ yd

17. 40 g

18. $14\dfrac{1}{16}$ mi

Chapter 3   DECIMALS

3.1 Introductions to Decimals
Mathematically Speaking
1. power
2. thousandths
3. increasing
4. right
5. hundredth

PRACTICE
1. a. tenths
   b. hundredths
   c. millionths

2. $\dfrac{5}{8}$

3. $2\dfrac{6}{25}$

4. a. nine hundred twenty-four thousandths
   b. four and seventy-three hundredths
   c. four tenths
5. a. 0.9
   b. 15.021
6. 0.226
7. a. 84.3
   b. 80
   c. 84.30
8. 207.61

Objective A
1. Six hundred twenty-nine thousandths

2. Forty-six and one thousand, two hundred forty-five ten thousandths
3. 17.0906
4. 243.52

Objective B

5. $\frac{7}{25}$

6. $1\frac{4}{25}$

7. $8\frac{31}{500}$

8. $23\frac{7}{20}$

Objective C

9. >
10. <
11. <
12. >
13. 0.07, 0.0718, 0.0725
14. 8.269, 8.27, 8.278

Objective D

15. 5.28
16. 4.0
17. 61.60
18. 89.101

Objective E

19. $3.77
20. More than
21. $29,900
22. $1,713,000

3.2 Adding and Subtracting Decimals
Mathematically Speaking

1. difference
2. decimal points
3. zeros
4. sum

PRACTICE

1. 3.509
2. 3.64
3. 8.959
4. 304.2 miles

Objective A

1. 206.67

2. 8.35752
3. 24.51
4. 4.32
5. 799.2
6. 3.21 inches
7. 3.58
8. 2.503
9. 24.493
10. 5.203
11. $7.52
12. 4.25
13. 43.5
14. 45.9

Objective B

15. $111.79
16. 14.7 million
17. 426.3 Kwh

3.3 Multiplying Decimals
Mathematically Speaking

1. three
2. five
3. factors
4. square

PRACTICE

1. 64.885
2. 68.068
3. $7.17

Objective A

1. 68.4
2. 2.66
3. 0.315
4. 64.885
5. 0.5856
6. 0.1835
7. 0.28768
8. 159.1
9. 384,700
10. 21.53
11. 0.5291
12. 0.85
13. $\approx 0.7 \times 0.05 = 0.035$
14. $\approx 600 \times 3 = 1800$

Objective B

15. $13.40
16. $18.69

17. $957.35

3.4 Dividing Decimals
Mathematically Speaking
1. repeating
2. two
3. quotient
4. decimal
PRACTICE
1. 3.25
2. 55.4
3. 22.78
4. 500 shares
Objective A
1. 0.375
2. 0.84
Objective B
3. 9.75
4. 4.5
5. 3.4
6. 0.583
7. 180
8. 1.45
9. 3.375
10. 38.5
11. 8.73
12. 5.64
13. $\approx 12 \div 0.4 = 30$
14. ≈ 60
Objective C
15. 14.7 Kwh/day
16. $4.69/lb
17. 19.5 miles/gal

Chapter 4
SOLVING SIMPLE EQUATIONS

4.1 Introduction to Basic Algebra
Mathematically Speaking
1. algebraic
2. constant
3. variable
4. evaluate
PRACTICE
1. a. 9 plus y

b. w less than 3
c. a divided by 2
2. a. 22
b. 5.4
c. 24
3. $(1,421 - p)$ points
Objective A (answers may vary)
1. x plus 10; The sum of x and 10
2. y minus 4; The difference between y and 4
3. The difference between 6 and z; 6 decreased by z
4. 2 times p; twice p
5. The product of $\frac{1}{2}$ and x; $\frac{1}{2}$ of x
6. y divided by 5; The quotient of y and 5
7. $x - 9$
8. $d + 15$
9. $2p$
10. $10 - x$
11. $6.2y$
12. $\dfrac{17}{n}$
Objective B
13. 5.7
14. 75.2
15. 7.3
16. 364.62
17. $\dfrac{1}{6}$
18. 60
Objective C
19. $2w$
20. $(48,629 - p)$ dollars
21. $(43.99 + m)$ dollars

4.2 Solving Addition and Subtraction Equations
Mathematically Speaking
1. translates
2. add
3. variable
4. equation

PRACTICE
1. a. $85 - w = 59$
    b. $y + 54 = 102$
2. a. $y = 22$
    b. $n = 18.1$
3. $187.90

Objective A
1. $x + 4 = 20$
2. $x - \dfrac{4}{3} = \dfrac{4}{15}$
3. $x + 7 = 36$
4. $x - 5\dfrac{3}{4} = 2\dfrac{3}{8}$

Objective B
5. 27
6. 8
7. 20
8. 13
9. 3.5
10. 12.9
11. $\dfrac{23}{3}$
12. $4\dfrac{11}{15}$

Objective C
13. $y - 23 = 87$; 110 yd
14. $w - 261 = 2636$; 2897 lb
15. $a + 3.4 = 96$; 92.6

4.3   Solving Multiplication and Division Equations
Mathematically Speaking
1. multiply
2. substituting
3. multiplication
4. equation
PRACTICE
1. a. $\dfrac{4.2}{a} = 1.3$
    b. $12 = \dfrac{1}{3}x$
2. a. 9
    b. 21

3. $30 = \dfrac{n}{4}$; 120 disks

Objective A
1. $10x = 65$
2. $\dfrac{y}{8} = 13$
3. $2x = 38$
4. $5a = 50$

Objective B
5. 7
6. 3
7. 32
8. 18
9. 30
10. $\dfrac{1}{3}$
11. 127.5
12. $\dfrac{7}{24}$

Objective C
13. $2c = 46$; 23 cans
14. $\dfrac{3}{4}g = 186$; 248 students
15. $308.59m = 11,109.24$; 36 months

Chapter 5
RATIO AND PROPORTION

5.1   Introduction to Ratios
Mathematically Speaking
1. denominator
2. like
3. quotient
4. simplest form
5. different
PRACTICE
1. 1:3
2. $\dfrac{18 \text{ pounds}}{15 \text{ weeks}} = \dfrac{6 \text{ pounds}}{5 \text{ weeks}}$
3. $\dfrac{350 \text{ copies}}{\$14} = \dfrac{25 \text{ copies}}{\$1}$
4. 573 mi/month

Objective A

1. $\dfrac{4}{5}$

2. $\dfrac{5}{3}$

3. $\dfrac{7}{400}$

4. $\dfrac{100}{11}$

5. $\dfrac{3}{10}$

6. $\dfrac{1}{1}$

Objective B

7. $\dfrac{8 \text{ feet}}{5 \text{ seconds}}$

8. $\dfrac{49 \text{ miles}}{2 \text{ hours}}$

9. $\dfrac{358 \text{ yards}}{3 \text{ games}}$

10. $\dfrac{2 \text{ teachers}}{19 \text{ students}}$

11. $\dfrac{1 \text{ can of paint}}{394 \text{ ft}^2}$

12. $\dfrac{\$280}{1 \text{ person}}$

13. 45 km/hr

14. 20 mi/gal

15. $15.50/hr

16. 1.8 oz/serving

17. $0.78/bar

18. $0.25/slice

Objective C

19. $0.15/min

20. $\dfrac{2}{1}$

21. $\dfrac{1.232 \text{ Euros}}{\$1 \text{ U.S.}}$

5.2 Solving Proportions
Mathematically Speaking

1. as
2. proportion
3. solve
4. cross products

PRACTICE

1. $w = 65$
2. 510 miles

Objective A

1. 12
2. 22
3. 14
4. 18
5. 100
6. $\dfrac{25}{2}$, or $12\dfrac{1}{2}$
7. $\dfrac{6}{7}$
8. $\dfrac{26}{9}$, or $2\dfrac{8}{9}$
9. 108
10. $\dfrac{75}{2}$, or $37\dfrac{1}{2}$

Objective B

11. 8 yrs
12. 180 million
13. $14\dfrac{1}{4}$ hr
14. 18 days
15. 12 gallons
16. 62.5 lb
17. $3150

Chapter 6   PERCENTS

6.1 Introduction to Percents
Mathematically Speaking

1. left
2. decimal
3. percent
4. fraction

PRACTICE

1. a. $\dfrac{3}{25}$

b. $\frac{7}{2} = 3\frac{1}{2}$

2. a. 0.55
   b. 0.05
3. a. 25.4%
   b. 12%
4. 24%
5. 70%

Objective A

1. $\frac{3}{25}$

2. $\frac{63}{100}$

3. $1\frac{1}{5}$

4. $\frac{1}{125}$

5. $\frac{1}{800}$

6. $\frac{33}{400}$

7. 0.46
8. 1.75
9. 0.725
10. 2
11. 0.0975
12. 1.147

Objective B

13. 8%
14. 37.5%
15. 172%
16. 0.25%
17. 75%
18. 28%

19. $44\frac{4}{9}\%$

20. 26%

Objective C

21. 0.276

22. $\frac{3}{10}$

23. 80%

24. $1\frac{1}{10}$

6.2 Solving Percent Problems
Mathematically Speaking

1. of
2. percent
3. amount
4. base

PRACTICE

1. a. 24
   b. 38.2%
   c. 240
2. $2,184

Objective A

1. 18
2. 525
3. $216
4. 0.14
5. 30
6. $150

7. $2\frac{2}{3}$

8. 40 miles
9. 15%
10. 175%
11. 25%
12. 85%

Objective B

13. 80%
14. 40%
15. 40 games

6.3 More on Percents
Mathematically Speaking

1. original
2. compound
3. on commission
4. discount

PRACTICE

1. 6.5%
2. 5.25%
3. $9,500
4. $715.75
5. $6,829.06

Objective A

1. $19\frac{3}{13}\%$

2. 7%

3. $8\frac{4}{7}\%$

4. 15%

5. 43.9%

Objective B

6. $10,850

7. $4,539

8. $116.88

9. $354.77

10. $422.21

11. Discount: $52.50; tax: $9.19; total cost: $131.69

Objective C

12. $272

13. $63

14. $3068

15. $5,650

16. $4,410

Chapter 7
SIGNED NUMBERS

7.1 Introduction to Signed Numbers
Mathematically Speaking

1. negative number
2. integers
3. absolute value
4. larger
5. opposites

PRACTICE

1.

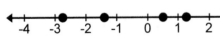

2. a. −15
    b. 87
3. a. positive; 21
    b. negative; 3.9
4. −1
5. $30

Objective A

1.

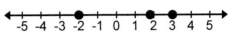

Objective B

2. −14

3. $3\frac{5}{9}$

Objective C

4. 16

5. 3.2

6. $\frac{9}{20}$

7. $2\frac{1}{6}$

Objective D

8. −2

9. 4

10. −3.4

11. $-4\frac{2}{5}$

Objective E

12. Randy's
13. Stacey
14. Tuesday
15. 1200 ft

7.2 Adding Signed Numbers
Mathematically Speaking

1. right
2. commutative
3. larger
4. absolute value

PRACTICE

1. −10
2. 6
3. 4°F

Objective A

1. −9
2. 7
3. 13
4. −33
5. 6.5
6. −3.6

7. $-\dfrac{23}{20}$

8. $\dfrac{19}{36}$

9. $-1\dfrac{7}{10}$

10. $-7\dfrac{5}{8}$

Objective B
11. $-12°F$
12. -410 employees
13. $-\$6,100$
14. $-\$34$
15. $13,796$ ft

7.3 Subtracting Signed Numbers
Mathematically Speaking
1. order of operations
2. addition
3. opposite
4. absolute value

PRACTICE
1. 5
2. 16
3. $-4\dfrac{1}{5}$
4. 13
5. $436.3°F$

Objective A
1. -24
2. -33
3. 20
4. 3.8
5. 58
6. -3.6
7. $\dfrac{7}{20}$
8. $-\dfrac{59}{36}$
9. $7\dfrac{31}{40}$
10. $-7\dfrac{4}{7}$
11. 32

12. -45
13. 118.4
14. 9

Objective B
15. 284,196 ft
16. 19,852 ft
17. $434.64
18. -0.07 sec

7.4 Multiplying Signed Numbers
Mathematically Speaking
1. even
2. positive
3. odd
4. negative

PRACTICE
1. -36
2. 0.84
3. $-\$645,000$

Objective A
1. -72
2. -105
3. 104
4. 600
5. $-1,800$
6. 0.42
7. 6.5
8. $-\dfrac{4}{9}$
9. -36
10. 0.09
11. 750
12. 24

Objective B
13. $-\$1,200$
14. $\$14,720$
15. $\$660$
16. $-3°F$

7.5 Dividing Signed Numbers
Mathematically Speaking
1. multiplication
2. equal
3. negative
4. positive

PRACTICE

1. 4

2. $-\dfrac{1}{3}$

3. -29 per day

Objective A

1. -3
2. -12
3. 50
4. -246
5. 54
6. -1.4
7. 60
8. $-\dfrac{7}{8}$
9. $\dfrac{16}{9}$
10. -20
11. $\dfrac{7}{9}$
12. $-\dfrac{2}{5}$
13. $-\dfrac{13}{3}$, or $-4\dfrac{1}{3}$
14. $\dfrac{19}{8}$, or $2\dfrac{3}{8}$

Objective B

15. 14.2 points per game
16. 3,300 people per year
17. 19 rounds
18. 8 batches

Chapter 8
BASIC STATISTICS

8.1 Introduction to Basic Statistics
Mathematically Speaking

1. statistics
2. range
3. mode
4. arithmetic
5. median

PRACTICE

1. mean: 6.875

median: 6.5
mode: 9
2. $15 - 4 = 11$
3. mean: \$69,600
median: \$78,000
mode: \$82,000
range: \$43,000

Objective A

1. mean: 371.8; median: 371;
mode: 371
2. mean: 33.25; median: 32;
mode: none
3. mean: \$40.07; median: \$42.31;
mode: \$48.36
4. mean: $5\dfrac{3}{5}$; median: $4\dfrac{1}{4}$;
mode: $4\dfrac{1}{4}$
5. mean: -4.5; median: -4;
mode: -4

Objective B

6. 49
7. 27
8. \$21.05
9. $6\dfrac{1}{3}$
10. 9

Objective C

11. 88 points/game
12. 71.2
13. Median: $79°$; mode: $75°$
14. 60
15. 2.727 or 2.73

8.2 Tables and Graphs
Mathematically Speaking

1. graph
2. rows
3. pictograph
4. histogram
5. circle graph

PRACTICE

1. a. 14 vehicles
   b. 30 vehicles

2. a. 5%
    b. $1290
Objective A
1. a. 3.5 million
    b. 11.4 million
2. a. 16 years
    b. 19 years
3. a. 7 more
    b. Kentucky and South Dakota
Objective B
4. a. Fares
    b. Customer Service
5. a. Malta
    b. About 850,000
6. a. 33%
    b. About 72
7. a. 5%
    b. $5.28 million
8. a. 82%
    b. Securities: $2,100; banks: $540; life insurance: $540; mutual funds: $2,820
9. a. 1985
    b. 2003
10. a. 60°F
    b. November
    c. About 48°F

Chapter 9
MORE ON ALGEBRA

9.1 Solving Equations
PRACTICE
1. $y = -5$
2. $n = -12$
3. $n = -6$
4. $y = -6$
5. $a = 32$
6. $45 + 35x = 167.50$; $x = 3.5$ hours
Objective A
1. 61
2. 194
3. -182
4. 32

5. $-\dfrac{4}{3}$
6. 8
7. $-\dfrac{32}{5}$
8. -144
Objective B
9. 6
10. -8
11. -11
12. 0
13. $\dfrac{5}{4}$
14. $\dfrac{20}{3}$
15. 6
16. 65
Objective C
17. $x + 85 = 415$; $330
18. $2x = 65,380$; $32,690
19. $22,000 - x = 18,750$; 3,250 ft
20. $20,000 - 1,600x = 800$; 12 years

9.2 More on Solving Equations
PRACTICE
1. $10x$
2. $t = 2$
3. $t = 9$
4. $n = -6$
Objective A
1. $11x$
2. $2y$
3. $12a + 5$
4. $7 + w$ or $w + 7$
Objective B
5. 6
6. 7
7. -6
8. 6
Objective C
9. -14
10. $\dfrac{44}{7}$

11. 12
12. 3

Objective D
13. $108
14. 50 m, 100 m, and 300 m
15. Length: 94 ft; width: 50 ft

9.3 Using Formulas
PRACTICE

1. Temperature $= \dfrac{c}{3} + 4$

2. 904 cubic inches
3. $A = 25$ square meters

Objective A

1. $A = \dfrac{1}{2}bh$

2. $P = 2l + 2w$
3. $C = \pi d$
4. $A = s^2$

5. $A = \dfrac{1}{2}\pi r^2$

Objective B
6. 15 ft^2
7. 25°C
8. $731.50
9. 8π cm ≈ 25.12 cm
10. 63 in^2

Objective C
13. 20π in.^{3} ≈ 62.83 in.^{3}
14. $2,164.86
15. 700 ft^3

Chapter 10
MEASUREMENT AND UNITS

10.1 U.S. Customary Units
Mathematically Speaking
1. unit
2. larger
3. length
4. gallon
5. numerator
PRACTICE

1. 108 feet
2. 12 ft 1 in
3. 10,560 feet

Objective A
1. 9 ft
2. 64 oz
3. 5,500 lb
4. 2.5 ft
5. 7 qt
6. 28 days
7. 28 pt
8. 3,200 oz
9. 2.5 day
10. 8 min 20 sec

Objective B
11. 2 lb 13 oz
12. 3 hr 35 min
13. 3 yr 4 mo
14. 2 ft 6 in.
15. 7 pt 8 fl oz, or 7 pt 1 c
16. a. 40 ounces
    b. 4.75 pounds
17. 32 ounces
18. 1,813 inches

10.2 Metric Units and Metric/U.S. Customary Unit Conversions
Mathematically Speaking
1. capacity
2. milli-
3. centi-
4. kilo-
PRACTICE
1. 3,200 km; length
2. 0.65
3. 700,000 mL

4. $3\dfrac{1}{3}$ L

5. 650 mL

Objective A
1. length
2. capacity
3. length
4. weight
5. capacity

Objective B
6. 8,000 mm
7. 2.7 L
8. 6,100 g
9. 1.3 g
10. 0.4875 cm
11. 5,000 ml

Objective C
12. 45 cm
13. 250 mi
14. 2.4 m
15. 1.6 lb
16. 19 L
17. 22.0 lb
18. 2.2 qt
19. 240 km

Objective D
20. a. 40 L
    b. about 10.5 gallons
21. about 281.25 miles
22. about 46.5 meters

Chapter 11
BASIC GEOMETRY

11.1 Introduction to Basic Geometry
Mathematically Speaking
1. line segment
2. supplementary
3. perpendicular
4. isosceles
5. scalene
6. diameter
7. obtuse
8. vertical
9. trapezoid
10. parallelogram

PRACTICE
1. $67°$
2. $52°$
3. 10 inches

Objective A
1. isosceles triangle
2. parallelogram
3. circle
4. trapezoid

5. $131.5°$
6. $153°$
7. $36°$
8. $m\angle 2 = 133°, \ m\angle 4 = 35°,$
   $m\angle 5 = 133°, \ m\angle 6 = 12°$

Objective B
9. $35°$
10. $113°$

Objective C
11. 1.15 inches
12. $72°$
13. $59°$

11.2 Perimeter and Circumference
Mathematically Speaking
1. circle
2. circumference
3. perimeter
4. composite

PRACTICE
1. 48 inches
2. about 43.96 meters
3. 128 inches
4. a. 384 yards
   b. $13,440

Objective A
1. 17.3 mm
2. $14\dfrac{1}{2}$ in.
3. 47
4. 48 ft
5. 18.84 yd
6. 75.36 ft
7. 25.12 in.
8. 43.96 in.

Objective B
9. 40.56 ft
10. 39.42 cm
11. 62 ft
12. 52.12 m

Objective C
13. 288 ft
14. 75.36 in.
15. 4605 ft

11.3   Area
Mathematically Speaking
1. triangle
2. trapezoid
3. square inches
4. circle

PRACTICE
1. 96 square inches
2. 153.86 square meters
3. 720 square inches
4. 324 square inches

Objective A
1. $40 \, \text{cm}^2$
2. $72 \, \text{ft}^2$
3. $200.96 \, \text{in}^2$
4. $11.31 \, \text{m}^2$
5. $5\frac{41}{64} \, \text{in}^2$

Objective B
6. $105.12 \, \text{ft}^2$
7. $57.87 \, \text{cm}^2$
8. $174 \, \text{ft}^2$
9. $57.76 \, \text{m}^2$
10. $23.44 \, \text{ft}^2$

Objective C
11. about $4.9 \, \text{cm}^2$
12. $44 \, \text{ft}^2$
13. $93.5 \, \text{in}^2$
14. square pizza: $196 \, \text{in}^2$; round pizza: $153.86 \, \text{in}^2$. The square pizza is the better buy.
15. $10,344.465 \, \text{m}^2$

11.4   Volume
Mathematically Speaking
1. volume
2. cylinder
3. composite
4. rectangular solid

PRACTICE
1. 125.6 cubic inches

2. 3,885.23 cubic feet
3. 78,500 cubic feet

Objective A
1. $409.2 \, \text{m}^3$
2. $3052.08 \, \text{cm}^3$
3. $1884 \, \text{m}^3$
4. $904.32 \, \text{ft}^3$
5. $540 \, \text{in}^3$

Objective B
6. $310.86 \, \text{in}^3$
7. $2,474.32 \, \text{cm}^3$
8. $717.4 \, \text{cm}^3$
9. $108 \, \text{ft}^3$
10. $651.36 \, \text{m}^3$

Objective C
11. $\frac{1}{64} \, \text{in}^3$
12. $330.55 \, \text{in}^3$
13. $9,110.19 \, \text{ft}^3$
14. $14.36 \, \text{m}^3$
15. $481\frac{7}{8} \, \text{ft}^3$

11.5   Similar triangles
Mathematically Speaking
1. corresponding
2. similar
3. in proportion
4. shape

PRACTICE
1. $y = 21$
2. 21 feet

Objective A
1. 18 m
2. 10 mi
3. 8.25 cm

Objective B
4. 36 ft
5. 69.6 ft
6. 105.6 cm
7. 84 ft
8. 52 feet

11.6 Square Roots and the Pythagorean Theorem
Mathematically Speaking
1. hypotenuse
2. perfect square
3. square root
4. Pythagorean Theorem

PRACTICE
1. 50
2. 8 and 9
3. $\sqrt{189} \approx 13.7$
4. $\sqrt{306} \approx 17.5$ feet

Objective A
1. 1
2. 5
3. 7
4. 12
5. 8 and 9
6. 2 and 3
7. 10 and 11
8. 6 and 7
9. 7.5
10. 4.6
11. 9.6
12. 13.2

Objective B
13. $c = 10$
14. $b = 9.8$
15. $b = 11.0$
16. $c = 4.2$

Objective C
17. 127.3 ft
18. 14.4 mi
19. 102.4 ft
20. 56 inch

Concept/Skill	Description	Example
[1.1] Place value		846,120
[1.1] To read a whole number	Working from left to right, • read the number in each period, and then • name the period in place of the comma.	71,400 is read "seventy-one thousand, four hundred".
[1.1] To write a whole number	Working from left to right, • write the number named in each period, and • replace each period name with a comma.	"Five thousand, twelve" is written 5,012.
[1.1] To round a whole number	• Underline the place to which you are rounding. • The digit to the right of the underlined digit is called the *critical digit*. Look at the critical digit—if it is 5 or more, add 1 to the underlined digit; if it is less than 5, leave the underlined digit unchanged. • Replace all the digits to the right of the underlined digit with zeros.	$386 \approx 390$ $4,817 \approx 4,800$
[1.2] Addend, sum	In an addition problem, the numbers being added are called *addends*. The result is called their *sum*.	$6 + 4 = 10$ Addend Addend Sum
[1.2] The identity property of addition	The sum of a number and zero is the original number.	$4 + 0 = 4$ $0 + 7 = 7$
[1.2] The commutative property of addition	Changing the order in which two numbers are added does not affect their sum.	$7 + 8 = 8 + 7$
[1.2] The associative property of addition	When adding three numbers, regrouping addends gives the same sum.	$(5 + 4) + 1 = 5 + (4 + 1)$
[1.2] To add whole numbers	• Write the addends vertically, lining up the place values. • Add the digits in the ones column, writing the right-most digit of the sum on the bottom. If the sum has two digits, carry the left digit to the top of the next column on the left. • Add the digits in the tens column as in the preceding step. • Repeat this process until you reach the last column on the left, writing the entire sum of that column on the bottom.	$\begin{array}{r} 7{,}385 \\ 92{,}551 \\ +2{,}007 \\ \hline 101{,}943 \end{array}$
[1.2] Minuend, subtrahend, difference	In a subtraction problem, the number that is being subtracted from is called the *minuend*. The number that is being subtracted is called the *subtrahend*. The answer is called the *difference*.	$10 - 6 = 4$ Minuend Subtrahend, Difference

Place value table:

Thousands			Ones		
Hundreds	Tens	Ones	Hundreds	Tens	Ones

4 is in the ten thousands place.

Concept/Skill	Description	Example

[1.2] To subtract whole numbers	• On top, write the number *from which* we are subtracting. On the bottom, write the number that is being *taken away*, lining up the place values. Subtract in each column separately.
	• Start with the ones column.
	   a. If the digit on top is *larger* than or *equal* to the digit on the bottom, subtract and write the difference below.
	   b. If the digit on top is *smaller* than the digit on the bottom, borrow from the digit to the left on top. Then subtract and write the difference below the bottom digit.
	• Repeat this process until the last column on the left is finished, subtracting and writing its difference below.
		$$\begin{array}{r} 8\ ^{1}4\ 1 \\ 7,\cancel{9}\ \cancel{5}\ 2 \\ -1,8\ 8\ 3 \\ \hline 6,0\ 6\ 9 \end{array}$$

[1.3] Factor, product	In a multiplication problem, the numbers being multiplied are called *factors*. The result is called their *product*.	Factor Product $4 \times 5 = 20$
[1.3] The identity property of multiplication	The product of any number and 1 is that number.	$1 \times 6 = 6$ $7 \times 1 = 7$
[1.3] The multiplication property of 0	The product of any number and 0 is 0.	$51 \times 0 = 0$
[1.3] The commutative property of multiplication	Changing the order in which two numbers are multiplied does not affect their product.	$3 \times 2 = 2 \times 3$
[1.3] The associative property of multiplication	When multiplying three numbers, regrouping the factors gives the same product.	$(4 \times 5) \times 6 = 4 \times (5 \times 6)$
[1.3] The distributive property	Multiplying a factor by the sum of two numbers gives the same result as multiplying the factor by each of the two numbers and then adding.	$2 \times (4 + 3)$ $\quad = (2 \times 4) + (2 \times 3)$
[1.3] To multiply whole numbers	• Multiply the top factor by the ones digit in the bottom factor and write this product. • Multiply the top factor by the tens digit in the bottom factor and write this product leftward, beginning with the tens column. • Repeat this process until all the digits in the bottom factor are used. • Add the partial products, writing this sum.	$$\begin{array}{r} 693 \\ \times\ \ 71 \\ \hline 693 \\ 48\ 51 \\ \hline 49{,}203 \end{array}$$
[1.4] Divisor, dividend, quotient	In a division problem, the number that is being used to divide another number is called the *divisor*. The number into which it is being divided is called the *dividend*. The result is called the *quotient*.	Quotient 3 $4\overline{)12}$ Divisor ⌐ ⌐ Dividend

Concept/Skill	Description	Example
[1.4] To divide whole numbers	• Divide 17 into 39, which gives 2. Multiply the 17 by 2 and subtract the result (34) from 39. Beside the difference (5), bring down the next digit (3) of the dividend. • Repeat this process, dividing the divisor (17) into 53. • At the end, there is a remainder of 2. Write it beside the quotient on top.	$$\begin{array}{r} 23\ R2 \\ 17\overline{)393} \\ \underline{34} \\ 53 \\ \underline{51} \\ 2 \end{array}$$
[1.5] Exponent (or power), base	An *exponent* (or *power*) is a number that indicates how many times another number (called the *base*) is used as a factor.	$\underset{\text{Base}}{\uparrow}$ Exponent$\downarrow$ $5^3 = 5 \times 5 \times 5$
[1.5] Order of operations rule	To evaluate mathematical expressions, carry out the operations *in the following order.* **1.** First, perform the operations within any grouping symbols, such as parentheses ( ) or brackets [ ]. **2.** Then, raise any number to its power $\blacksquare^{\text{※}}$. **3.** Next, perform all multiplications and divisions as they appear from left to right. **4.** Finally, do all additions and subtractions as they appear from left to right. 	( )
$\blacksquare^{\text{※}}$
$\times \quad \div$
$+ \quad -$		$\begin{aligned} 8 + 5 \cdot (3 + 1)^2 &= 8 + 5 \cdot 4^2 \\ &= 8 + 5 \cdot 16 \\ &= 8 + 80 \\ &= 88 \end{aligned}$
[1.5] Average (or mean)	The *average* (or *mean*) of a set of numbers is the sum of those numbers divided by however many numbers are in the set.	The average of 3, 4, 10, and 3 is 5 because $$\frac{3 + 4 + 10 + 3}{4} = \frac{20}{4} = 5$$
[1.6] To solve word problems	• Read the problem carefully. • Choose a strategy (such as drawing a picture, breaking up the question, substituting simpler numbers, or making a table). • Decide which basic operation(s) are relevant and then translate the words into mathematical symbols. • Perform the operations. • Check the solution to see if the answer is reasonable. If it is not, start again by rereading the problem.	

Concept/Skill	Description	Example
[2.1] Prime number	A whole number that has exactly two different factors: itself and 1.	2, 3, 5
[2.1] Composite number	A whole number that has more than two factors.	4, 8, 9
[2.1] Prime factorization of a whole number	The number written as the product of its prime factors.	$30 = 2 \cdot 3 \cdot 5$
[2.1] Least common multiple (LCM) of two or more whole numbers	The smallest nonzero whole number that is a multiple of each number.	The LCM of 30 and 45 is 90.
[2.1] To compute the least common multiple (LCM)	• Find the prime factorization of each number. • Identify the prime factors that appear in each factorization. • Multiply these prime factors, using each factor the greatest number of times that it occurs in any of the factorizations.	$20 = 2 \cdot 2 \cdot 5$ $= 2^2 \cdot 5$ $30 = 2 \cdot 3 \cdot 5$ The LCM of 20 and 30 is $2^2 \cdot 3 \cdot 5$, or 60.
[2.2] Fraction	Any number that can be written in the form $\frac{a}{b}$, where a and b are whole numbers and b is nonzero.	$\frac{3}{11}, \frac{9}{5}$
[2.2] Proper fraction	A fraction whose numerator is smaller than its denominator.	$\frac{2}{7}, \frac{1}{2}$
[2.2] Mixed number	A number with a whole-number part and a proper fraction part.	$5\frac{1}{3}, 4\frac{5}{6}$
[2.2] Improper fraction	A fraction whose numerator is greater than or equal to its denominator.	$\frac{9}{4}, \frac{5}{5}$
[2.2] To change a mixed number to an improper fraction	• Multiply the denominator of the fraction by the whole-number part of the mixed number. • Add the numerator of the fraction to this product. • Write this sum over the denominator to form the improper fraction.	$4\frac{2}{3} = \frac{3 \times 4 + 2}{3}$ $= \frac{14}{3}$
[2.2] To change an improper fraction to a mixed number	• Divide the numerator by the denominator. • If there is a remainder, write it over the denominator.	$\frac{14}{3} = 4\frac{2}{3}$
[2.2] To find an equivalent fraction	Multiply the numerator and denominator of $\frac{a}{b}$ by the same whole number; that is, $\frac{a}{b} = \frac{a \cdot n}{b \cdot n}$, where both b and n are nonzero.	$\frac{3}{4} = \frac{3 \cdot 2}{4 \cdot 2} = \frac{6}{8}$

Concept/Skill	Description	Example
[2.2] To simplify a fraction	• First, express both the numerator and denominator as the product of their prime factors. • Then, divide out or cancel all common factors.	$\dfrac{30}{84} = \dfrac{\overset{1}{\cancel{2}} \cdot \overset{1}{\cancel{3}} \cdot 5}{\underset{1}{\cancel{2}} \cdot 2 \cdot \underset{1}{\cancel{3}} \cdot 7} = \dfrac{5}{14}$
[2.2] Like fractions	Fractions with the same denominator.	$\dfrac{2}{5}, \dfrac{3}{5}$
[2.2] Unlike fractions	Fractions with different denominators.	$\dfrac{3}{5}, \dfrac{3}{10}$
[2.2] To compare fractions	• If the fractions are like, compare their numerators. • If the fractions are unlike, write them as equivalent fractions with the same denominator and then compare their numerators.	$\dfrac{6}{8}, \dfrac{7}{8}$ $6 < 7$, so $\dfrac{6}{8} < \dfrac{7}{8}$ $\dfrac{2}{3}, \dfrac{12}{15}$ or $\dfrac{10}{15}, \dfrac{12}{15}$ $12 > 10$, so $\dfrac{12}{15} > \dfrac{2}{3}$
[2.2] Least common denominator (LCD) of two or more fractions	The least common multiple of their denominators.	The LCD of $\dfrac{11}{30}$ and $\dfrac{7}{45}$ is 90.
[2.3] To add (or subtract) like fractions	• Add (or subtract) the numerators. • Use the given denominator. • Write the answer in simplest form.	$\dfrac{1}{8} + \dfrac{1}{8} = \dfrac{2}{8} = \dfrac{1}{4}$ $\dfrac{3}{8} - \dfrac{1}{8} = \dfrac{2}{8} = \dfrac{1}{4}$
[2.3] To add (or subtract) unlike fractions	• Write the fractions as equivalent fractions with the same denominator, usually the LCD. • Add (or subtract) the numerators, keeping the same denominator. • Write the answer in simplest form.	$\dfrac{2}{3} + \dfrac{1}{2} = \dfrac{4}{6} + \dfrac{3}{6}$ $= \dfrac{7}{6}$, or $1\dfrac{1}{6}$ $\dfrac{5}{12} - \dfrac{1}{6} = \dfrac{5}{12} - \dfrac{2}{12}$ $= \dfrac{3}{12}$, or $\dfrac{1}{4}$
[2.3] To add mixed numbers	• Write the fractions as equivalent fractions with the same denominator, usually the LCD. • Add the fractions. • Add the whole numbers. • Write the answer in simplest form.	$\begin{aligned} 4\dfrac{1}{2} &= 4\dfrac{3}{6} \\ +6\dfrac{2}{3} &= +6\dfrac{4}{6} \\ \hline &10\dfrac{7}{6} = 11\dfrac{1}{6} \end{aligned}$
[2.3] To subtract mixed numbers	• Write the fractions as equivalent fractions with the same denominator, usually the LCD. • Regroup or borrow from the whole number on top if the fraction on the bottom is larger than the fraction on top. • Subtract the fractions. • Subtract the whole numbers. • Write the answer in simplest form.	$\begin{aligned} 4\dfrac{1}{5} &= 3\dfrac{6}{5} \\ -1\dfrac{2}{5} &= -1\dfrac{2}{5} \\ \hline &2\dfrac{4}{5} \end{aligned}$

Concept/Skill	Description	Example
[2.4] **To multiply fractions**	• Multiply the numerators. • Multiply the denominators. • Write the answer in simplest form.	$\dfrac{1}{2}\cdot\dfrac{3}{5}=\dfrac{3}{10}$
[2.4] **To multiply mixed numbers**	• Write the mixed numbers as improper fractions. • Multiply the fractions. • Write the answer in simplest form.	$2\dfrac{1}{2}\cdot1\dfrac{2}{3}=\dfrac{5}{2}\cdot\dfrac{5}{3}$ $=\dfrac{25}{6},\text{ or }4\dfrac{1}{6}$
[2.4] **Reciprocal of** $\dfrac{a}{b}$	The fraction $\dfrac{b}{a}$ formed by switching the numerator and denominator.	The reciprocal of $\dfrac{4}{3}$ is $\dfrac{3}{4}$.
[2.4] **To divide fractions**	• Change the divisor to its reciprocal, and multiply the resulting fractions. • Write the answer in simplest form.	$\dfrac{2}{5}\div\dfrac{3}{7}=\dfrac{2}{5}\cdot\dfrac{7}{3}$ $=\dfrac{14}{15}$
[2.4] **To divide mixed numbers**	• Write the mixed numbers as improper fractions. • Divide the fractions. • Write the answer in simplest form.	$2\dfrac{1}{2}\div1\dfrac{1}{3}=$ $\dfrac{5}{2}\div\dfrac{4}{3}=$ $\dfrac{5}{2}\cdot\dfrac{3}{4}=\dfrac{15}{8},\text{ or }1\dfrac{7}{8}$

Concept/Skill	Description	Example
[3.1] Decimal	A number written with two parts: a whole number, which precedes the decimal point, and a fractional part, which follows the decimal point.	Whole-number part → Fractional part → 3.721 ↑ Decimal point
[3.1] Decimal place	A place to the right of the decimal point.	Decimal places ↓ 8.035 ↑↑↑ Tenths Thousandths Hundredths
[3.1] To change a decimal to the equivalent fraction or mixed number	• Copy the nonzero whole-number part of the decimal and drop the decimal point. • Place the fractional part of the decimal in the numerator of the equivalent fraction. • Make the denominator of the equivalent fraction 1 followed by as many zeros as the decimal has decimal places. • Simplify the resulting fraction, if possible.	The decimal 3.25 is equivalent to the mixed number $3\dfrac{25}{100}$ or $3\dfrac{1}{4}$.
[3.1] To compare decimals	• Rewrite the numbers vertically, lining up the decimal points. • Working from left to right, compare the digits that have the same place value. At the first place value where the digits differ, the decimal which has the largest digit with this place value is the largest decimal.	1.073 1.06999 In the ones place and the tenths place, the digits are the same. But in the hundredths place, $7 > 6$, so $1.073 > 1.06999$.
[3.1] To round a decimal to a given decimal place	• Underline the digit in the place to which the number is being rounded. • The digit to the right of the underlined digit is called the *critical digit*. Look at the critical digit—if it is 5 or more, add 1 to the underlined digit; if it is less than 5, leave the underlined digit unchanged. • Drop all decimal places to the right of the underlined digit.	$23.9381 \approx 23.94$ ↑ Critical digit
[3.2] To add decimals	• Rewrite the numbers vertically, lining up the decimal points. • Add. • Insert a decimal point in the sum below the other decimal points.	0.035 0.08 C+ 0.00813 0.12313
[3.2] To subtract decimals	• Rewrite the numbers vertically, lining up the decimal points. • Subtract, inserting extra zeros in the minuend if necessary for regrouping. • Insert a decimal point in the difference below the other decimal points.	0.90370 C− 0.17052 0.73318

Concept/Skill	Description	Example
[3.3] To multiply decimals	• Multiply the factors as if they were whole numbers. • Find the total number of decimal places in the factors. • Count that many places from the right end of the product, and insert a decimal point.	21.07 ← Two decimal places $\times\ \ 0.18$ ← Two decimal places 3.7926 ← Four decimal places
[3.4] To change a fraction to the equivalent decimal	• Divide the denominator of the fraction into the numerator, inserting to its right both a decimal point and enough zeros to get an answer either without a remainder or rounded to a given decimal place. • Place a decimal point in the quotient directly above the decimal point in the dividend.	$\dfrac{7}{8} = 8\overline{)7.000}^{\,0.875}$
[3.4] To divide decimals	• Move the decimal point in the divisor to the right end of the number. • Move the decimal point in the dividend the same number of places to the right as in the divisor. • Insert a decimal point in the quotient directly above the decimal point in the dividend. • Divide the new dividend by the new divisor, inserting zeros at the right end of the dividend as necessary.	$3.5\overline{)71.05} =$ $\begin{array}{r} 20.3 \\ 35\overline{)710.5} \\ \underline{70} \\ 10 \\ \underline{0} \\ 10.5 \\ \underline{10.5} \end{array}$

Concept/Skill	Description	Example
[4.1] Variable	A letter that represents an unknown number.	x, y, t
[4.1] Constant	A known number.	$2, \frac{1}{3}, 5.6$
[4.1] Algebraic expression	An expression that combines variables, constants, and arithmetic operations.	$x + 3, \frac{1}{8}n$
[4.1] To evaluate an algebraic expression	• Substitute the given value for each variable. • Carry out the computation.	Evaluate $8 - x$ for $x = 3.5$: $8 - x = 8 - 3.5$, or 4.5
[4.2] Equation	A mathematical statement that two expressions are equal.	$2 + 4 = 6, x + 5 = 7$
[4.2] To solve addition or subtraction equations	• For an addition equation, subtract the same number from each side of the equation in order to isolate the variable on one side.	$y + 9 = 15$ $y + 9 - 9 = 15 - 9$ $y = 6$ Check $\quad y + 9 = 15$ $6 + 9 \overset{?}{=} 15$ $15 \overset{\checkmark}{=} 15$
	• For a subtraction equation, add the same number to each side of the equation in order to isolate the variable on one side.	$w - 6\frac{1}{2} = 8$ $w - 6\frac{1}{2} + 6\frac{1}{2} = 8 + 6\frac{1}{2}$ $w = 14\frac{1}{2}$
	• In either case, check the solution by substituting the value of the unknown in the original equation to verify that the resulting equation is true.	Check $\quad w - 6\frac{1}{2} = 8$ $14\frac{1}{2} - 6\frac{1}{2} \overset{?}{=} 8$ $8 \overset{\checkmark}{=} 8$
[4.3] To solve multiplication or division equations	• For a multiplication equation, divide by the same number on each side of the equation in order to isolate the variable on one side.	$1.3r = 26$ $\dfrac{1.3r}{1.3} = \dfrac{26}{1.3}$ $r = 20$ Check $\quad 1.3r = 26$ $1.3(20) \overset{?}{=} 26$ $26 \overset{\checkmark}{=} 26$
	• For a division equation, multiply by the same number on each side of the equation in order to isolate the variable on one side.	$\dfrac{x}{7} = 8$ $7 \cdot \dfrac{x}{7} = 7 \cdot 8$ $x = 56$
	• In either case, check the solution by substituting the value of the unknown in the original equation to verify that the resulting equation is true.	Check $\quad \dfrac{x}{7} = 8$ $\dfrac{56}{7} \overset{?}{=} 8$ $8 \overset{\checkmark}{=} 8$

Concept/Skill	Description	Example
[5.1] Ratio	A comparison of two quantities expressed as a quotient.	3 to 4, $\frac{3}{4}$, or 3:4
[5.1] Rate	A ratio of unlike quantities.	$\dfrac{10 \text{ students}}{3 \text{ tutors}}$
[5.1] To simplify a ratio	• Write the ratio as a fraction. • Express the fraction in simplest form. • If the quantities are alike, drop the units. If the quantities are unlike, keep the units.	9:27 is the same as 1:3, because $\dfrac{9}{27} = \dfrac{1}{3}$ 21 hours to 56 hours $= \dfrac{21 \text{ hours}}{56 \text{ hours}} = \dfrac{21}{56} = \dfrac{3}{8}$ 175 miles per 7 gallons $= \dfrac{175 \text{ miles}}{7 \text{ gallons}} = \dfrac{25 \text{ miles}}{1 \text{ gallon}}$, or 25 mpg
[5.1] Unit rate	A rate in which the number in the denominator is 1.	$\dfrac{180 \text{ calories}}{1 \text{ ounce}}$, or 180 calories per ounce, or 180 cal/oz
[5.1] Unit price	The price of one item, or one unit.	$0.69 per can, or $0.69/can
[5.2] Proportion	A statement that two ratios are equal.	$\dfrac{5}{8} = \dfrac{15}{24}$
[5.2] To solve a proportion	• Find the cross products, and set them equal. • Solve the resulting equation. • Check the solution by substituting the value of the unknown in the original equation to verify that the resulting proportion is true.	$\dfrac{6}{9} = \dfrac{2}{x}$ $6x = 18$ $x = 3$ Check $\dfrac{6}{9} = \dfrac{2}{x}$ $\dfrac{6}{9} \stackrel{?}{=} \dfrac{2}{3}$ $6 \cdot 3 \stackrel{?}{=} 9 \cdot 2$ $18 \stackrel{\checkmark}{=} 18$

Concept/Skill	Description	Example
[6.1] **Percent**	A ratio or fraction with denominator 100. It is written with the % sign, which means divided by 100.	$7\% = \dfrac{7}{100}$ ↑ Percent
[6.1] **To change a percent to the equivalent fraction**	• Drop the % sign from the given percent and place the number over 100. • Simplify the resulting fraction, if possible.	$25\% = \dfrac{25}{100} = \dfrac{1}{4}$
[6.1] **To change a percent to the equivalent decimal**	• Drop the % sign from the given percent and divide the number by 100.	$23.5\% = .235,\ \text{or } 0.235$
[6.1] **To change a decimal to the equivalent percent**	• Multiply the number by 100 and insert a % sign.	$0.125 = 12.5\%$
[6.1] **To change a fraction to the equivalent percent**	• Multiply the fraction by 100%.	$\dfrac{1}{5} = \dfrac{1}{5} \times 100\% = \dfrac{1}{\cancel{5}} \times \dfrac{\overset{20}{\cancel{100}}}{1}\%$ $= 20\%$
[6.2] **Base**	The number that we are taking the percent of. It always follows the word *of* in the statement of a percent problem.	50% of 8 is 4. ↑ Base
[6.2] **Amount**	The result of taking the percent of the base.	50% of 8 is 4. ↑ Amount
[6.2] **To solve a percent problem using the translation method**	• Translate as follows: What number, what percent → x is → = of → × or · % → decimal or fraction • Set up the equation. **The percent of the base is the amount.** • Solve.	What is 50% of 8? $x = 0.5 \cdot 8$ $x = 4$ 30% of what number is 6? $0.3 \cdot x = 6$ $\dfrac{\cancel{0.3}x}{\cancel{0.3}} = \dfrac{6}{0.3}$ $x = \dfrac{6}{0.3} = 20$ What percent of 8 is 2? $x \cdot 8 = 2$ $8x = 2$ $x = \dfrac{2}{8} = \dfrac{1}{4} = 25\%$

Concept/Skill	Description	Example
[6.2] To solve a percent problem using the proportion method	• Identify the amount, the base, and the percent, if known. • Set up and substitute into the proportion. $$\frac{\text{Amount}}{\text{Base}} = \frac{\text{Percent}}{100}$$ • Solve for the unknown quantity.	50% of 8 is what number? $$\frac{x}{8} = \frac{50}{100}$$ $$100x = 400$$ $$x = 4$$ 30% of what number is 6? $$\frac{6}{x} = \frac{30}{100}$$ $$30x = 600$$ $$x = 20$$ What percent of 8 is 2? $$\frac{2}{8} = \frac{x}{100}$$ $$8x = 200$$ $$x = 25$$ So the answer is 25%.
[6.3] To find a percent increase or decrease	• Compute the difference between the new and the original values. • Determine what percent this difference is of the original value.	Find the percent increase for a quantity that changes from 4 to 5. Difference: $5 - 4 = 1$ What percent of 4 is 1? x · 4 = 1 $$x = \frac{1}{4} = 0.25, \text{ or } 25\%$$

Concept/Skill	Description	Example
[7.1] Positive number	A number greater than 0.	$5, \frac{1}{3}, 2.7$
[7.1] Negative number	A number less than 0.	$-5, -\frac{1}{3}, -2.7$
[7.1] Signed number	A number with a sign that is either positive or negative.	$5, -5, \frac{1}{3}, -\frac{1}{3}, 2.7, -2.7$
[7.1] Integers	The numbers $\ldots, -4, -3, -2, -1, 0, 1, 2, 3, 4, \ldots$ continuing indefinitely in both directions.	$+5, -5$
[7.1] Opposites	Two numbers that are the same distance from 0 on the number line but on opposite sides of 0.	$+2$ and -2
[7.1] Absolute value	The distance of a number from 0 on the number line, represented by the symbol $\mid\ \mid$.	2 units ⌐ ⌐2 units $\begin{array}{c}\xleftrightarrow{\ \ \ \ \ \ \ \ \ \ \ \ \ \ \ } \\ -4\ -3\ -2\ -1\ \ 0\ \ 1\ \ 2\ \ 3\ \ 4\end{array}$ $\mid -2\mid = 2, \quad \mid +2\mid = 2$
[7.1] To compare signed numbers	• Locate the points being compared on the number line. A number to the right is larger than a number to the left.	$\begin{array}{c}\xleftrightarrow{\ \ \ \ \ \ \ \ \ \ \ \ \ \ \ } \\ -4\ -3\ -2\ -1\ \ 0\ \ 1\ \ 2\ \ 3\ \ 4\end{array}$ $2 > -1$
[7.2] To add two signed numbers	• If the numbers have the same sign, add the absolute values and keep the sign. • If the numbers have different signs, subtract the smaller absolute value from the larger and take the sign of the number with the larger absolute value.	$-0.5 + (-1.7) = -2.2$ because $\mid -0.5\mid + \mid -1.7\mid =$ $\qquad 0.5 + 1.7 = 2.2$ $3\frac{1}{2} + (-9) = -5\frac{1}{2}$ because $\mid -9\mid > \left\lvert +3\frac{1}{2}\right\rvert$ and $9 - 3\frac{1}{2} = 5\frac{1}{2}$
[7.3] To subtract two signed numbers	• Change the operation of subtraction to addition. Then add the first number and the opposite of the second number. • Follow the rule for adding signed numbers.	$\begin{aligned}-2 - (-5) &= -2 + 5 \\ &= +3, \text{ or } 3\end{aligned}$
[7.4] To multiply two signed numbers	• Multiply the absolute values of the numbers. • If the numbers have the same sign, their product is positive; if the numbers have different signs, their product is negative.	$(-8)\left(-\frac{1}{2}\right) = +4, \text{ or } 4$ $-0.2 \times 4 = -0.8$
[7.5] To divide two signed numbers	• Divide the absolute values of the numbers. • If the numbers have the same sign, their quotient is positive; if the numbers have different signs, their quotient is negative.	$\frac{-8}{-4} = +2, \text{ or } 2$ $18 \div (-2) = -9$

Concept/Skill	Description	Example
[8.1] Mean	Given a set of numbers, the sum of the numbers divided by however many numbers are in the set.	For 0, 0, 1, 3, and 5, the mean is: $$\frac{0+0+1+3+5}{5}$$ $$=\frac{9}{5}=1.8$$
[8.1] Median	Given a set of numbers arranged in numerical order, the number in the middle. If there are two numbers in the middle, the mean of the two middle numbers.	For 0, 0, 1, 3, and 5, the median is 1.
[8.1] Mode	Given a set of numbers, the number (or numbers) occurring most frequently in the set.	For 0, 0, 1, 3, and 5, the mode is 0.
[8.1] Range	Given a set of numbers, the difference between the largest and the smallest number in the set of numbers.	For 0, 0, 1, 3, and 5, the range is $5 - 0$, or 5.
[8.2] Table	A rectangular display of data.	
[8.2] Pictograph	A graph in which images such as people, books, or coins are used to represent the quantities.	
[8.2] Bar graph	A graph in which quantities are represented by thin, parallel rectangles called bars. The length of each bar is proportional to the quantity that it represents.	
[8.2] Histogram	A graph of a frequency table.	

Concept/Skill	Description	Example
[8.2] Line graph	A graph in which points are connected by straight-line segments. The position of any point on a line graph is read against the vertical axis and the horizontal axis.	
[8.2] Circle graph	A graph that resembles a pie (a whole amount) that has been cut into slices (the parts).	

Page 391

Concept/Skill	Description	Example
[9.1] Solution	A value of the variable that makes an equation a true statement	-2 is a solution of $x + 5 = 3$ because $-2 + 5 = 3$.
[9.1] To solve an equation with more than one operation	• First, use the rule for solving addition or subtraction equations. • Then, use the rule for solving multiplication or division equations.	$2y - 7 = 13$ $2y - 7 + 7 = 13 + 7$ $2y = 20$ $\dfrac{2y}{2} = \dfrac{20}{2}$ $y = 10$
[9.2] Like terms	Terms that have the same variables with the same exponents	$2x$ and $3x$
[9.2] Unlike terms	Terms that are not like	$4a^2$ and $3a$
[9.2] To combine like terms	• Use the distributive property. • Add or subtract.	$2x + 3x = (2 + 3)x$ $= 5x$
[9.3] Formula	An equation that indicates how variables are related to one another	$I = \dfrac{V}{R}$

U.S. Customary Units		Relationships
[10.1] Length	Mile (mi), yard (yd), foot (ft), and inch (in.)	5,280 ft = 1 mi 3 ft = 1 yd 12 in. = 1 ft
[10.1] Weight	Ton, pound (lb), and ounce (oz)	2,000 lb = 1 ton 16 oz = 1 lb
[10.1] Capacity (liquid volume)	Gallon (gal), quart (qt), pint (pt), cup (c), and fluid ounce (fl oz)	4 qt = 1 gal 2 pt = 1 qt 2 c = 1 pt 8 fl oz = 1 c
[10.1] Time	Year (yr), month (mo), week (wk), day, hour (hr), minute (min), and second (sec)	365 days = 1 yr 12 mo = 1 yr 52 wk = 1 yr 7 days = 1 wk 24 hr = 1 day 60 min = 1 hr 60 sec = 1 min

METRIC PREFIXES

Prefix	Symbol	Meaning
Kilo-	k	One thousand (1,000)
Hecto-	h	Hundred (100)
Deka-	da	Ten (10)
Deci-	d	One tenth $\left(\dfrac{1}{10}\right)$
Centi-	c	One hundredth $\left(\dfrac{1}{100}\right)$
Milli-	m	One thousandth $\left(\dfrac{1}{1,000}\right)$

Metric Units		Relationships
[10.2] Length	Kilometer (km), meter (m), centimeter (cm), and millimeter (mm)	1,000 m = 1 km 100 cm = 1 m 1,000 mm = 1 m
[10.2] Weight	Kilogram (kg), gram (g), and milligram (mg)	1,000 g = 1 kg 1,000 mg = 1 g
[10.2] Capacity (liquid volume)	Kiloliter (kL), liter (L), and milliliter (mL)	1,000 L = 1 kL 1,000 mL = 1 L
[10.2] Time	Same as U.S. customary units.	

OTHER METRIC PREFIXES

Prefix	Symbol	Meaning
Giga-	G	One billion (1,000,000,000)
Mega-	M	One million (1,000,000)
Micro-	mc	One millionth $\left(\dfrac{1}{1,000,000}\right)$

OTHER METRIC UNITS

Unit	Symbol	Meaning	Relationship
Gigabyte	GB	1,000,000,000 bytes	1,000,000,000 B $=$ 1 GB
Megabyte	MB	1,000,000 bytes	1,000,000 B $=$ 1 MB
Microsecond	mcsec	$\dfrac{1}{1,000,000}$ second	1,000,000 mcsec $=$ 1 sec
Microgram	mcg	$\dfrac{1}{1,000,000}$ gram	1,000,000 mcg $=$ 1 g

Key Metric/U.S. Unit Conversion Relationships

[10.2] **Length**	1.6 km $\approx$ 1 mi 1,600 m $\approx$ 1 mi 3,300 ft $\approx$ 1 km 3.3 ft $\approx$ 1 m 39 in. $\approx$ 1 m 30 cm $\approx$ 1 ft 2.5 cm $\approx$ 1 in.
[10.2] **Weight**	910 kg $\approx$ 1 ton 2.2 lb $\approx$ 1 kg 450 g $\approx$ 1 lb 28 g $\approx$ 1 oz
[10.2] **Capacity (liquid volume)**	260 gal $\approx$ 1 kL 3.8 L $\approx$ 1 gal 1.1 qt $\approx$ 1 L 2.1 pt $\approx$ 1 L 470 mL $\approx$ 1 pt

Concept/Skill	Description	Example
[11.1] **Point**	An exact location in space, with no dimension.	•A
[11.1] **Line**	A collection of points along a straight path, that extends endlessly in both directions.	$\overleftrightarrow{AB}$
[11.1] **Line segment**	A part of a line having two endpoints.	$\overline{BC}$
[11.1] **Ray**	A part of a line having only one endpoint.	$\overrightarrow{AB}$
[11.1] **Angle**	Two rays that have a common endpoint called the *vertex* of the angle.	$\angle ABC$
[11.1] **Plane**	A flat surface that extends endlessly in all directions.	
[11.1] **Straight angle**	An angle whose measure is 180°.	180°
[11.1] **Right angle**	An angle whose measure is 90°.	
[11.1] **Acute angle**	An angle whose measure is less than 90°.	65°
[11.1] **Obtuse angle**	An angle whose measure is more than 90° and less than 180°.	120°
[11.1] **Complementary angles**	Two angles the sum of whose measures is 90°.	25° 65°
[11.1] **Supplementary angles**	Two angles the sum of whose measures is 180°.	40° 140°

Concept/Skill	Description	Example
[11.1] Intersecting lines	Two lines that cross.	
[11.1] Parallel lines	Two lines on the same plane that do not intersect.	$\overleftrightarrow{EF} \parallel \overleftrightarrow{GH}$
[11.1] Perpendicular lines	Two lines that intersect to form right angles.	$\overleftrightarrow{RT} \perp \overleftrightarrow{PQ}$
[11.1] Vertical angles	Two opposite angles with equal measure formed by two intersecting lines.	
[11.1] Polygon	A closed plane figure made up of line segments.	
[11.1] Triangle	A polygon with three sides.	
[11.1] Quadrilateral	A polygon with four sides.	
[11.1] Equilateral triangle	A triangle with three sides equal in length.	$\overline{PQ}$, $\overline{QR}$, and $\overline{PR}$ have equal lengths.
[11.1] Isosceles triangle	A triangle with two or more sides equal in length.	$\overline{AB}$ and $\overline{BC}$ have equal lengths.

Concept/Skill	Description	Example
[11.1] Scalene triangle	A triangle with no sides equal in length.	$\overline{HG}$, $\overline{GI}$, and $\overline{HI}$ have unequal lengths.
[11.1] Acute triangle	A triangle with three acute angles.	
[11.1] Right triangle	A triangle with one right angle.	
[11.1] Obtuse triangle	A triangle with one obtuse angle.	
[11.1] The sum of the measures of the angles of a triangle	In any triangle, the sum of the measures of all three angles is 180°.	$m\angle A + m\angle B + m\angle C = 180°$
[11.1] Trapezoid	A quadrilateral with only one pair of opposite sides parallel.	$\overrightarrow{AB} \parallel \overrightarrow{CD}$
[11.1] Parallelogram	A quadrilateral with both pairs of opposite sides parallel. Opposite sides are equal in length, and opposite angles have equal measures.	$\overline{LM} \parallel \overline{PO}$ $\overline{LP} \parallel \overline{MO}$ $\overline{LM}$ and $\overline{PO}$ have equal lengths, and $\overline{LP}$ and $\overline{MO}$ have equal lengths.
[11.1] Rectangle	A parallelogram with four right angles.	

Concept/Skill	Description	Example
[11.1] Square	A rectangle with four sides equal in length.	*D* *E* *F* *G* $\overline{DE}$, $\overline{EG}$, $\overline{FG}$, and $\overline{DF}$ have equal lengths.
[11.1] The sum of the measures of the angles of a quadrilateral	In any quadrilateral, the sum of the measures of the angles is 360°.	*B* *C* *A* *D* $m\angle A + m\angle B + m\angle C + m\angle D = 360°$
[11.1] Circle	A closed plane figure made up of points that are all the same distance from a fixed point called the center.	*O*
[11.1] Diameter	A line segment that passes through the center of a circle and has both endpoints on the circle.	*A* •*O* *B* Diameter $\overline{AB}$
[11.1] Radius	A line segment with one endpoint on the circle and the other at the center.	•*O* *B* Radius $\overline{OB}$
[11.2] Perimeter	The distance around a polygon.	7 cm 3 cm 2 cm 6 cm 5 cm $P = 3 + 7 + 2 + 5 + 6 = 23$ $P = 23$ cm
[11.2] Circumference	The distance around a circle.	10 in. $C = 10\pi \approx 31.4$ $C \approx 31.4$ in.

Concept/Skill	Description	Example
[11.3] Area	The number of square units that a figure contains.	$A = 4 \cdot 3 = 12$ $A = 12 \text{ in}^2$
[11.4] Volume	The number of cubic units required to fill a three-dimensional figure.	$V = 2 \cdot 3 \cdot 4 = 24$ $V = 24 \text{ in}^3$
[11.5] Similar triangles	Triangles that have the same shape but not necessarily the same size.	
[11.5] Corresponding sides	In similar triangles, the sides opposite the equal angles.	In the similar triangles pictured, $\overline{AB}$ corresponds to $\overline{DE}$, $\overline{BC}$ corresponds to $\overline{EF}$, and $\overline{AC}$ corresponds to $\overline{DF}$.
[11.5] To find a missing side of a similar triangle	• Write the ratios of the lengths of the corresponding sides. • Write a proportion using a ratio with known terms and a ratio with an unknown term. • Solve the proportion for the unknown term.	$\triangle TRS \sim \triangle XYW$ Find a. $\dfrac{ST}{WX} = \dfrac{TR}{XY}$ $\dfrac{4}{6} = \dfrac{8}{a}$ $4a = 48$ $a = 12$, or 12 in.
[11.6] Perfect square	A number that is the square of a whole number.	49 and 144
[11.6] (Principal) square root of n	The positive number, written $\sqrt{n}$, whose square is n.	$\sqrt{36}$ and $\sqrt{8}$

Concept/Skill	Description	Example
[11.6] Pythagorean theorem	For every right triangle, the sum of the squares of the lengths of the two legs equals the square of the length of the hypotenuse, that is, $$a^2 + b^2 = c^2$$ where a and b are the lengths of the legs, and c is the length of the hypotenuse.	Find a. $$a^2 + b^2 = c^2$$ $$a^2 + (\mathbf{24})^2 = (25)^2$$ $$a^2 + 576 = 625$$ $$a^2 + 576 - 576 = 625 - 576$$ $$a^2 = 49$$ $$a = \sqrt{49}$$ $$= 7, \text{ or } 7 \text{ yd}$$

Key Formulas

Figure	Formula	Example
[11.2]–[11.3] Triangle	*Perimeter* $$P = a + b + c$$ Perimeter equals the sum of the lengths of the three sides. *Area* $$A = \frac{1}{2}bh$$ Area equals one-half the base times the height.	$$P = a + b + c$$ $$= \mathbf{6} + 10 + 8$$ $$= 24, \text{ or } 24 \text{ m}$$ $$A = \frac{1}{2}bh$$ $$= \frac{1}{\underset{1}{2}} \cdot \overset{5}{\mathbf{10}} \cdot 4.8$$ $$= 24, \text{ or } 24 \text{ m}^2$$
[11.2]–[11.3] Rectangle	*Perimeter* $$P = 2l + 2w$$ Perimeter equals twice the length plus twice the width. *Area* $$A = lw$$ Area equals the length times the width.	$$P = 2l + 2w$$ $$= 2(\mathbf{7}) + 2(3)$$ $$= 14 + 6$$ $$= 20, \text{ or } 20 \text{ in.}$$ $$A = lw$$ $$= \mathbf{7} \cdot 3$$ $$= 21, \text{ or } 21 \text{ in}^2$$

Figure	Formula	Example
[11.2]–[11.3] Square	*Perimeter* $$P = 4s$$ Perimeter equals four times the length of a side. *Area* $$A = s^2$$ Area equals the square of a side.	$\frac{1}{2}$ in. $P = 4s$ $= 4 \cdot \dfrac{1}{2}$ $= 2$, or 2 in. $A = s^2$ $= \left(\dfrac{1}{2}\right)^2$ $= \dfrac{1}{4}$, or $\dfrac{1}{4}$ in^2
[11.3] Parallelogram	*Area* $$A = bh$$ Area equals the base times the height.	3 ft 6 ft $A = bh$ $= 6 \cdot 3$ $= 18$, or 18 ft^2
[11.3] Trapezoid	*Area* $$A = \frac{1}{2}h(b + B)$$ Area equals one-half the height times the sum of the bases.	3 in. 4 in. 5 in. $A = \dfrac{1}{2}h(b + B)$ $= \dfrac{1}{2} \cdot 4\,(3 + 5)$ $= \dfrac{1}{\overset{2}{\underset{1}{\cancel{2}}}} \cdot \overset{2}{\cancel{4}} \cdot 8$ $= 16$, or 16 in^2
[11.2]–[11.3] Circle	*Circumference* $$C = \pi d, \text{ or } C = 2\pi r$$ Circumference equals π times the diameter, or 2 times π times the radius. *Area* $$A = \pi r^2$$ Area equals π times the square of the radius.	8 cm $C = \pi d$ $\approx 3.14(8)$ ≈ 25.12, or 25.12 cm $A = \pi r^2$ $\approx 3.14\,(4)^2$ $\approx 3.14\,(16)$ ≈ 50.24, or 50.24 cm^2 Note: $d = 8$ cm, so $r = 4$ cm.

Figure	Formula	Example
[11.4] Rectangular solid	*Volume* $$V = lwh$$ Volume equals length times width times height.	 5 cm 7 cm 15 cm $V = lwh$ $= \mathbf{15} \cdot \mathbf{7} \cdot \mathbf{5}$ $= 525$, or 525 cm^3
[11.4] Cube	*Volume* $$V = e^3$$ Volume equals the cube of the edge.	 2 in. $V = e^3$ $= (\mathbf{2})^3$ $= 2 \cdot 2 \cdot 2$ $= 8$, or 8 in^3
[11.4] Cylinder	*Volume* $$V = \pi r^2 h$$ Volume equals π times the square of the radius times the height.	 12 m 4 m $V = \pi r^2 h$ $\approx 3.14\,(\mathbf{4})^2(\mathbf{12})$ $\approx 3.14\,(16)(12)$ ≈ 603, or 603 m^3
[11.4] Sphere	*Volume* $$V = \frac{4}{3}\pi r^3$$ Volume equals $\frac{4}{3}$ times π times the cube of the radius.	 2 ft $V = \frac{4}{3}\pi r^3$ $\approx \frac{4}{3}(3.14)(\mathbf{2})^3$ $\approx \frac{4}{3}(3.14)(8)$ $\approx \frac{100.48}{3}$ ≈ 33, or 33 ft^3